POLLUTION SCIENCE, TECHNOLOGY AND ABATEMENT

NOISE POLLUTION

SOURCES, EFFECTS ON WORKPLACE PRODUCTIVITY AND HEALTH IMPLICATIONS

POLLUTION SCIENCE, TECHNOLOGY AND ABATEMENT

Additional books in this series can be found on Nova's website under the Series tab.

Additional e-books in this series can be found on Nova's website under the e-book tab.

POLLUTION SCIENCE, TECHNOLOGY AND ABATEMENT

NOISE POLLUTION

SOURCES, EFFECTS ON WORKPLACE PRODUCTIVITY AND HEALTH IMPLICATIONS

EUGENE ROBERTO NICCHI
EDITOR

nova publishers
New York

For permission to use material from this book please contact us:
Telephone 631-231-7269; Fax 631-231-8175
Web Site: http://www.novapublishers.com

NOTICE TO THE READER

Library of Congress Cataloging-in-Publication Data

Noise pollution : sources, effects on workplace productivity and health implications / editor, Eugene Roberto Nicchi.
pages cm. -- (Pollution science, technology and abatement)
Includes index.
ISBN 978-1-63321-110-0 (hardcover)
1. Offices--Noise. 2. Noise--Health aspects. 3. Labor productivity. I. Nicchi, Eugene Roberto.
TD893.6.O34N65 2014
363.74--dc23
2014018886

Published by Nova Science Publishers, Inc. † New York

CONTENTS

PREFACE

Environmental noise is defined as the noise emitted from sources such as road, rail and air traffic, industries, construction and public works, and the neighborhood. Noise has been ranked high among forms of pollutions, which include air pollution, radioactive waste pollution, water pollutions, etc. In recent years, noise has received considerable worldwide attention as a result of the many studies linking noise pollution to various health effects that include auditory as well non-auditory health effects. It should be noted here that the effect of noise is seldom catastrophic, and is often only transitory. However, its adverse effects can be cumulative with prolonged or repeated exposure. Noise with daily activities and it causes sleep disruption, masking of speech, reduction in performance and the inability to enjoy one's property or leisure time. It impairs the quality of life. In addition, several studies have linked extended exposure to high noise levels to cardiovascular diseases, including high blood pressure and irregular heart beat. It has also been reported that high noise levels have an economic effects, especially tourism and real estate sectors, where it has been reported that the prices of houses tend to be higher in quite areas. There is also some evidence that noise can adversely affect general health and well-being in the same manner as chronic stress. Because of these factors, and the recent public awareness of the adverse effects of noise in the last few decades, many studies and noise monitoring programs have been established in various parts in the world. These studies and programs focus on measurements of perceived sound levels in communities, in the workplace (occupational noise), near airports and in public parks. A second area of research, which is getting more attention in recent years, focuses on assessing the effects of noise on individuals and communities, i.e. noise annoyance. This book discusses several topics that include driver style influence in the vehicle acoustic emissions in urban traffic; reaction to hybrid noise in communication; otoacoustic emissions by product distortion in metallurgical workers exposed to different doses of occupational noise; underwater noise pollution; and noise pollution in Zimbabwe.

Chapter 1 - Nightingales (*Luscinia megarhynchos* L.) are an excellent bio-indicator of environmental change in cities. Many studies show an interesting relation between their song and noise pollution. In fact this species can modify its acoustic spectra by traffic noise by adapting their dominant frequency. However no studies have made a complete analysis of wild and urban Nightingales studying a high number of sonometric characteristics in a enough high sample size of song records. In this paper the variability between both groups of birds had been determined using 34 bioacoustic parameters, formant and song structure and spectral characteristics. Using different mathematical procedures the authors determine that nightingales increase dominant frequency with urban noise pollution, decreasing simultaneously the intensity of the call. However, the authors do not observe changes in the number of formants, total energy, local peaks or other spectral characteristics of this bird species. Urban nightingales reduce their spectral entropy, the number of silent periods between songs and the number and time of territorial calls (rapid broadband trills). In contrast, this bird species increases the frequency and evenness of their songs. The author's results suggest a great acoustic plasticity in this species due to in the urban environment the communication is masked by low-frequency background noise, the abundance of males is lower and their acoustic variability decreases. The species, although migrant, returns to the places where it was born. Consequently, in the long term it is expected that these processes will have evolutionary effects.

Chapter 2 - The attenuation of traffic noise in highly populated urban environments is one of the biggest technological challenges that the scientific community has to face. Although the legislation has gradually imposed more severe limitations to the levels emitted by vehicles, these restrictions have not been sufficient to create a sustainable acoustic environment due to the constant increase in the car fleet and because the approval test (Directive 70/157/EEC) does not represent properly the noise level emitted by a car in real driving conditions.

The European Commission has placed this issue within the priority research strategies. Following these guidelines and recommendations, this work presents a new strategy for the control of car noise emissions, and at the same time improves the understanding of the fundamental parameters that influence in the car noise emissions.

Nowadays there are no studies about the influence on global acoustic contamination due to a vehicle as individual source neither about the influence of driving style. Because of this, it is impossible to identify which drivers are responsible for generating the maximum noise levels within the traffic stream.

Results show that there is a very clear correlation between these highest levels of noise and the nuisance generated, so concrete actions for the control of traffic noise should be directed towards noisier vehicles, rather than imposing limitations on the levels produced by the total traffic stream. So, it is necessary to develop studies to correlate radiated noise and driving behaviour.

In this work, a sample of vehicles has been instrumented to obtain physical measurements of the variables that can influence on the noise emission level. Vehicle positions, vehicle speed, longitudinal accelerations and drive time have been analyzed using a GPS sensor. Engine speed, engine load, throttle position and gear shift position have been studied through the vehicle CAN BUS and a battery of microphones has measured emitted noise in several points of the vehicle.

A circuit that represents the usual circulation in a big city like Madrid has been defined, as well as several vehicle models with great presence in existing fleet of cars. Several drivers have covered the circuit at different times of the day and previous parameters have been measured in each test in order to determine average values of behavior.

To conclude, previous data have been analyzed and noise equivalent levels have been identified with different test configurations. With these average values, it has been possible to define the noise emissions produced by a vehicle in real driving conditions and to identify the noisier driving style.

As results of the investigation, it has been demonstrated that a noisy driving style produces up to 9 dB (A) more on average than other drivers. Depending on the engine type a vehicle can emit about 5 dB more of engine noise at same drive style. There is no appreciable influence of the driver experience over the considered driving interest parameters. There is no influence of the driver sex over the considered driving interest parameters.

Chapter 3 - The noise pollution is a vast subject. Noise can be controlled and much of the technology for solving most noise problems is available today. To control the noise, there are many noise measurements, monitoring devices and techniques. However, the noise pollution and its effects on the environment is not really improved. The authors are interested in how far noise pollution disturbs communication and what can be done. Here, the authors provide three different case studies where they discuss how an effective communication can place in a noisy environment. The authors consider three different types of noise such as mild, time varying steady-unsteady and strong. As situations the authors considered experimental laboratory, health care support, and diagnostic support for persons with possible dementia. In this chapter, the authors describe how a human and a machine can communicate in these three scenarios in order to make the workplace and resources more productive.

Chapter 4 - Risks arising from physical agents - such as noise pollution - are sometimes underrated or not treated in a suitable way. This happens especially when workplace wellbeing is considered. In Europe, security is ensured with the *Directive 2003/10/EC of the European Parliament and of the Council of February 6th, 2003 on the minimum health and safety requirements regarding the exposure of workers to the risks arising from physical agents (noise)*. It supplies the minimum health requirements to assure health-care and protection from loud sounds to those workers who are exposed to noise as a result of their work. Nevertheless, this standard is far away from guaranteeing a complete safeguard because of several shortcomings.

In this chapter, the working activities of workers who remain in noisy workplaces for prolonged periods are analyzed. These activities may produce serious risks for worker's health with particular reference to the organ of hearing and its psychophysics functions. Therefore, parameters and acoustics quantities - which allow a better understanding of the physical phenomenon appointed within the Directive - are analyzed using data recorded with dosimeters and real time analyzers. Works carried out in environments characterized by different acoustic intensities are considered in this analysis. Depending on exposure time, each noisy environment can produces annoyance phenomena or permanent psychosomatic damage. In particular, the following activities are subject to investigation: office activities, work activities of public transportation drivers, university and research center's laboratories (e.g. mechanics labs, engine testing room, material stress testing labs, etc.) and industrial activities.

Based on the sampled data, acoustics parameters used within the current Directive are evaluated to highlight that it underestimates the effective noise to which workers are subjected. The same data is used to prove that these parameters cannot guarantee a protection from noise as effective as in the Directive's original aims. In practice, through several in-situ measurements the intensity of acoustic parameters is evaluated according to what is stated within the Directive. Then, the same data are analyzed using an integrative procedure – proposed by the authors – to evaluate the influence of impulsive components, which cannot be determined with the current standards. In addition, the influence of sound phenomena's spectral composition and its influence on the interested portion of human cochlea are achieved.

Chapter 5 - Introduction: The otoacoustic emissions test is useful in monitoring changes of hearing not yet detectable by tonal audiometry, as well as to monitor the damage in the cochlea caused by exposure to noise. Objective: To investigate the evoked otoacoustic emissions by distortion-product in metallurgical workers exposed to different doses of occupational noise. Method: This was a prospective cohort study conducted at normal hearing metallurgical workers. The evaluated subjects were divided in three groups: not exposed to noise (Group I), sporadically exposed to noise (Group II) and often exposed to occupational noise (group III). Results: There were significant changes in otoacoustic emissions in Groups II and III, considering both ears. In the amplitude and signal/noise ratio it was observed that the higher the frequency of the sound, the worse the results in Groups II and III, and the greater the exposure to noise, the lower the average found in otoacoustic emissions. The frequency with the worst records was 6 kHz. The Group II, although sporadically exposed to noise, suffered the effects of exposure and revealed a major risk for hearing loss. Conclusion: Evoked otoacoustic emissions were shown to be a sensitive method to detect changes in cochlear function in both groups.

Chapter 6 - The use of personal ear protection is the only immediate solution for high noise level until a noise control action is carried out.

In this chapter the noise leak pathways through the protector to the ear were investigated.

Measurements with artificial head were performed to evaluate the influence of various parameters on the sound attenuation provided by hearing protectors. The parameters evaluated were: 1) Effect of circumaural cushion, 2) Effect of cup internal lining, 3) Effect of cup weight, 4) Effect of headband force, and the 5) Effect of cup vibration.

The results showed that: 1) Coherence Function between cup vibration and sound pressure inside the cup has shown that higher sound energy is transmitted, at low frequencies, due to cup vibrations. Vibration measured axially at the cup outer surface shows mechanical resonances started around 150 Hz; 2) Internal foam lining of the cup is an important component of the earmuffs to avoid acoustic resonances of the cup cavity; 3) The attenuation measured with a complete earmuff and the same without the cushion has shown that the acoustic energy leakage occurs at low frequency bands. This leakage is due to the low mass density of the cushion materials - foam in this case -; 4). As the headband force increases to a maximal deformation of the cushion, attenuation increases, especially at low frequencies. The increase in the headband force is however a limiting factor to the use of the earmuff, as it can get too uncomfortable; 5) Higher attenuation is obtained at low frequencies as the cup weight increases. This is easily explained by the acoustic transmission mass law. Note that as the protector weight increase, other factors, such as comfort, decreases.

The use of hearing protectors is very important to protect workers against high sound pressure levels in the workplace. On the other hand, also very important, is the quality and maintenance of these protection devices. In the case of earmuffs, it was demonstrated that the removal of the internal foam lining generates acoustic resonances of the cup cavity, which substantially reduces the attenuation provided by the protector at high frequencies.

Chapter 7 - Today, offshore wind power is a small but growing industry because wind yields offshore are greater than yields for onshore installation. Average wind speeds over water are typically 20% higher than at nearby onshore locations. The most significant concerns relating to wind farms appear to center on noise production. However, (i) only a few studies deal with the impact from offshore wind farms on fish, (ii) knowledge of noise effects on marine life is limited, and (iii) there are significant data gaps. Thus, data on the impact of underwater noise from offshore wind farms on marine life have not yet been compiled in a systematic manner. It therefore seems worthwhile to draw a general picture concerning the effect of underwater noise from wind power turbines on marine life. The following general picture is drawn in this chapter: the short-term potential impact during pre-construction; the short-term intensive impact during construction; and the physiological and/or masking effects that may occur over a long period while the wind farm is in operation. It can be concluded that offshore wind farms should be listed in noise databases in order to promote rational environment management.

Chapter 8 - Noise control is essential in an enclosed machine room where high levels of reverberant and direct sound are encountered. Because pure tone noise is harmful to workers and may lead to psychological and physiological ailments, a traditional Helmholtz resonator that reduces pure tone noise has been commonly employed. However, the conventional Helmholtz resonator which is equipped with an externally extended tube requires a larger space. For existing plants that are in need of sufficient operation and maintenance space, a smaller designed Helmholtz resonator is necessary. In order to overcome the above drawbacks of an oversized Helmholtz resonator, a modified Helmholtz resonator that is compact and equipped with an internally extended tube is proposed. To appreciate the response of resonance frequencies and sound absorption coefficients with respect to various acoustical mechanisms and design parameters, an experimental study that varies the design parameters (L_k, d_k, F_k, H) is performed. Additionally, a two-layer resonator connected by two one-layer resonators in series is also discussed.

Results reveal that the equivalent shape factor (F_k) will increase when the ratio of the length to diameter for the resonating tube decreases. Both the resonance frequency (f_{res}) and related sound absorption coefficient (*abs*) will increase when the d_k increases. In addition, the resonance frequency (f_{res}) will decrease and the related sound absorption coefficient (*abs*) will increase when the H increases. Subsequently, both the resonance frequency (f_{res}) and sound absorption coefficient (*abs*) will decrease simultaneously when the L_k increases.

Consequently, the resonance frequency (f_{res}) of the two-layer modified Helmholtz resonator (2MHR-001 and 2MHR-002) will be tuned at higher frequencies than that of the one-layer resonator (1MHR030 and 1MHR040).

Chapter 9 - Previous research on noise levels arising form residential road traffic in the City of Sharjah, United Arab Emirates (UAE) has shown that noise levels in the city is high, exceeding national as well as international noise limits and standards. These studies have linked noise exposure in residential-industrial mixed area with annoyance, which is an indicator of more serious, chronic health conditions. In this chapter the authors report the results of a study conducted recently to explore the association between residential road traffic noise and self-reported annoyance for residents of the City of Sharjah. A comprehensive survey was conducted on randomly selected subjects aged 19-59 years, residing minimum of 5 years in the area and residing within 50 m of the roadways. The survey questions included two parts, the first of which contained demographic questions, while the second part contained questions aimed at gauging annoyance levels encountered by residents as a result of exposure to high noise levels. The results of the study showed that more than 67% of subjects feel that their neighborhood is noisy. To explore the association between noise levels and self-reported annoyance, noise levels were classified in units of 5 dB (A) from 65 dB (A) to 80 dB (A). Self-reported annoyance was marked at levels above 65-70 dB (A). Gender dependence was evident, where the association was found to be statistically significant for female subjects and the prevalence of annoyance was more for male subjects. Both age and period of residence were found to be significant predictors of annoyance. The results of this study further suggest the association residential noise exposure near roadways and self-reported annoyance, among the study subjects.

Chapter 10 - Noise pollution is a growing problem in Zimbabwe and other developing countries especially those in sub-Saharan Africa. It impairs people's hearing abilities, damages their health and can lead to death. This chapter examines noise pollution in Zimbabwe with special reference to sources, effects and health implications as well as work place productivity. It is based on primary and secondary sources of information. A questionnaire survey was conducted in December, 2013 in order to determine various aspects of the problem in both rural and urban areas of Zimbabwe. It was preceded by literature review on the subject matter. Some of the issues discussed include: the country's legislation on noise pollution, causes and effects of noise in urban and rural areas as well as suggested solutions on the problems and challenges exposed. It is noted that noise pollution is more pronounced in urban than rural areas as indicated by the number of victims.

The chapter ends with an examination of noise pollution in Harare's furniture manufacturing sector. It shows that levels of pollution in most cases exceed those prescribed by the country's laws on noise pollution. Consequently, solutions should be found in order to address the problem at local and national levels. However, these solutions should be based on research if they have to be effective. Since Zimbabwe is a cash strapped economy, it lacks the capacity to fund such expensive research projects. There is a need for the country to solicit for foreign donations so as to conduct researches on noise pollution issues. However, before this can be done, the country will have to improve its relations with some Western donor countries, such as Britain and the USA.

In: Noise Pollution
Editor: Eugene Roberto Nicchi

ISBN: 978-1-63321-110-0

Chapter 1

SONOMETRIC VARIABILITY OF URBAN AND WILD NIGHTINGALES (*LUSCINIA MEGARHYNCHOS* L.)

Daniel Patón*[1], *A. Jiménez and A. Hurtado
Numerical Ecology. Ecology Unit, Department of Plant Biology, Ecology and Earth Sciences, Faculty of Sciences, University of Extremadura, Badajoz, Spain

ABSTRACT

Nightingales (*Luscinia megarhynchos* L.) are an excellent bio-indicator of environmental change in cities. Many studies show an interesting relation between their song and noise pollution. In fact this species can modify its acoustic spectra by traffic noise by adapting their dominant frequency. However no studies have made a complete analysis of wild and urban Nightingales studying a high number of sonometric characteristics in a enough high sample size of song records. In this paper the variability between both groups of birds had been determined using 34 bioacoustic parameters, formant and song structure and spectral characteristics. Using different mathematical procedures we determine that nightingales increase dominant frequency with urban noise pollution, decreasing simultaneously the intensity of the call. However, we do not

[1] E-mail: dpaton@unex.es.

observe changes in the number of formants, total energy, local peaks or other spectral characteristics of this bird species. Urban nightingales reduce their spectral entropy, the number of silent periods between songs and the number and time of territorial calls (rapid broadband trills). In contrast, this bird species increases the frequency and evenness of their songs. Our results suggest a great acoustic plasticity in this species due to in the urban environment the communication is masked by low-frequency background noise, the abundance of males is lower and their acoustic variability decreases. The species, although migrant, returns to the places where it was born. Consequently, in the long term it is expected that these processes will have evolutionary effects.

1. INTRODUCTION

In 2050, 68% of the world's population will live in cities (United Nations, 2011). This increase in urbanization is causing permanent changes in diversity of avian assemblages by forest fragmentation (Husté et al., 2006). Consequently, bird communities in urban areas are being increasingly studied by their bioindicator values and urban gardens are now considered potential ecological reserves (Patón et al., 2013). We are witnessing a very recent natural experiment very recent in evolutionary terms: the adaptation of wildlife to the urban ecosystem (Ortega, 2012). Urbanization causes new challenges in biological conservation, because it forces us to redefine the concept of nature reserves. In fact, in developed countries it is no longer possible to think of vast natural reserves without human intervention. Moreover, the urbanites demand to know the pristine wild nature that can be seen in the TV documentaries and cinemas but also want to have something of that nature near where they live. This necessity transforms urban green areas into perfect places where these natural interactions can be studied, to realize programs of environmental education or include artificial reserves of fauna and flora. In addition, urban areas are interesting in many aspects of biological conservation but also like theoretical and experimental models where the laws of ecology can be are studied (Patón et al., 2012). In fact, we can consider the areas of parks and gardens as ecological islands or metapopulations separated by a matrix of buildings (Fernández-Juricic et al., 2001). This fact allows the study of the effect of different environmental processes on these small environments. Issues such as the relationship between diversity and park area, the influence of structural aspects of gardens on wildlife richness, changes in breeding ecology or urban birds, differences in foraging behavior, the increase in longevity, the decrease in animal health or the effect of ecotones in the animal behavior can be easily studied in urban parks and gardens (Luniak, 2004). For example, it is possible to separate the forest specialists inside urban parks from generalist species on the edges (Fernández-Juricic et al., 2001). Particularly interesting is the study of the environmental impacts that affect wildlife in urban areas. More specifically, the effect of noise pollution on small birds is receiving increasing attention (Patón et al., 2012). This is because many birds present similar hearing systems to humans and can be considerred as perfect animal models (Brumm, 2006).

For many wildlife species, urban areas are impassable barriers. However, other species can adapt to these artificial environments taking advantage of the reduced presence of predators and the large food and water supplies of urban parks (Jokimäki and Huhta, 2000). Most birds that adapt to urban environments in Mediterranean Europe are highly common, present a cosmopolitan geographic distribution and a generalist feeding behavior. This is the case of the House Sparrow (Passser domesticus L.), Rock Dove (Columba livia Gmelin) or Spotless Starling (*Sturnus unicolor* Temminck). However, other species are forest specialists that live inside urban parks and gardens where the effect of pedestrians and noise pollution is lower. Among these species is the Common Nightingale (*Luscinea megarhynchos* Brehm), a small Eurasian insectivore and migrating bird that inhabits the scrublands areas of the temperate forests. This species is an perfect example of the adaptability of acoustic behavior to changing environmental conditions. Due to the complexity of their song, the Nightingale has been extensively studied. Certain studies indicate that nightingales learn their songs due to a balance between conespecific signals and environmental conditions (Hultsch et al., 1999). As with other birds, the songs of Nightingales have the double purpose of attracting females and rejecting other males (Schmidt et al., 2008). In fact, the volume of the Nightingale's song is a good indicator of their corporal condition and its relation to nutritional levels in the early stages of development (Hultsch et al., 1999). Also, the complexity of trills can indicate the behavioral adaptability of males. Males with a more complex song can be more attractive to females by their singularity and also are more clearly differentiated from adjacent males of other territories (Sprau et al., 2012). Playback experiments had shown that this bird has different parts in its songs that vary according environmental conditions, especially due to density of other males and agonistic encounters. These parts are initial whistles with low amplitude (6-8 kHz), normal whistles, normal calls that are more complex, trills that are very fast broadband calls and silent periods (Sprau et al., 2010). In situations of social stress, nightingales increase initial whistles and broadband trills and decrease the frequency and length of silences (Sprau et al., 2010). Consequently, nightingales use a considerable quantity of sound energy in certain environmental situations. Some studies have shown that urban birds present larger brains than wild specimens of the same species (Maklakov et al., 2011). This is an evolutionary effect caused by the high variability in the environmental conditions of urban areas and their strong effects on the adaptability of small birds. This adaptability is directly related with intelligence in animals. Although the nightingale is a migratory wild species it can be found in urban areas where it adapt its song to these noisy environments (Hultsch et al., 1999). In this sense, certain studies show that Nightingales increase their song amplitude as adaptation to high levels of noise pollution caused mainly by traffic and pedestrians (Brumm, 2004). Another effect that causes this high background noise in cities is the decrease in the hearing distance of the Nightingale's songs. Males cannot attract females or defend their territories so efficiently as in natural conditions. Consequently, this effect has severe consequences on reproduction success. In this sense, some studies have found that Nightingales change their acoustic parameters according to the size of their territories or males densities (Sprau et al., 2012). In areas, where these birds are in high densities certain agonistic calls named rapid broadband trills increase considerably (Naguib and Todt, 1997). In this sense, we had found densities of Nightingales in urban areas of SW Spain very close to maximum densities observed in natural environments. For all these considerationsfactors, we can consider Nightingales to be perfect bioindicators of the possible effect of noise pollution in wild birds. However, we cannot forget that this species does not permanently reside in the cities, so any observed change must be considered mainly as a temporal behavioral adaptation and non-effect of natural selection except in males that systematically return to urban parks where they were born. This is a process of a great theoretical interest, since it determines the degree in that urban noise difficulty the communication of these birds and the possible evolutionary effects in the long-termwhich main communication difficulties the urban noise in wild birds produces and the possible evolutionary effects in the long-term. In this paper, we explore in depth various acoustic parameters comparing Nightingale songs of many individuals in urban and wild areas. Finally we made theoretical inferences about the main consequences that urban noise could have in the long-term on this wild bird and how these processes can explain the avian assemblages of urban areas.

2. MATERIAL AND METHODS

Audio files of a total set of 100 nightingales were obtained with a MARANTZ PMD 661 sound recorder connected to a Parabolic Microphone Telinga 4-PRO during spring of the years 2011, 2012 and 2013. The audio format was 24 bits uncompressed and stereo that is the maximum quality that this recorder permits. The sampling areas were located in the surroundings and parks of the city of Cáceres (Extremadura, SW Spain), forest rivers of the Monfragüe National Park (Extremadura, SW Spain) and the villages of "Sotillo de la Adrada" (Avila, Central Spain) and Mogarraz (Salamanca, Central Spain) (Figure 1). In this sampling design, two types of areas with very clear differences in the degree of urbanization were defined. On the one hand, natural forest areas (n = 61 birds) with no, or rather weak human presence and on the other urban areas (n = 39 birds) with human presence and influence of traffic and pedestrian noise. Within these urban areas urban parks, gardens and places close to highways or roads were selected. Areas with an intermediate degree of urbanism such as rural, exurban or suburban places were not sampled in order to determine, the two extreme conditions of the rural gradient.

Figure 1. Locations were Nightingale's songs had been sampled in Central-West Spain. CAC: Cáceres city, MNP: Monfragüe National Park (Cáceres province), MOG: Mogarraz (Salamanca province), STA: Sotillo de la Adrada village (Avila province).

Table 1. Description of 34 sound parameters of Nightingales and the software used in the present paper

Parameter	Description	Software
ACI	Acoustic complexity index	Seewave
ADI	Acoustic diversity index	Soundecology
AEI	Acoustic evenness index	Soundecology
BI	Bioacoustic index	Soundecology
CENT	Centroid	Seewave
dB	Decibel	Praat
E	Energy in Pascal^2 s^{-1}	Praat
H	Acoustic total entropy	Seewave
IQR	Inter-quartile range	Seewave
KURT	Kurtosis of spectra	Seewave
MEA	Mean frequency	Seewave
MED	Median frequency	Seewave
MOD	Mode frequency	Seewave
NLP	Number of local peaks of normalized spectra	Audacity
PREC	Frequency precision	Seewave
Q25	First quartile of frequency	Seewave
Q75	Third quartile of frequency	Seewave
RSS	Ratio between songs and silents	Seewave
S	Total songs by minute	Seewave
SDM	Standard deviation of the mean frequency	Seewave
SEM	Standard error of the mean frequency	Seewave
SFM	Spectral flatness	Seewave
SG%N	Percentage of the number of normal songs	Audacity
SG%T	Percentage of the time of normal songs	Audacity
SGNT	Number of normal songs by minute	Audacity
SH	Spectral entropy	Seewave
SKEW	Skewness of spectra	Seewave
SL%N	Percentage of the number of silents	Audacity
SL%T	Percentage of the time of silents	Audacity
SLNT	Number of silents by minute	Audacity
TH	Entropy of a temporal envelope	Seewave
TR%N	Percentage of the number of rapid broadband trills	Audacity
TR%T	Percentage of the time of rapid broadband trills	Audacity
TRNT	Number of rapid broadband trills by minute	Audacity

The background noise of wind, water, traffic, pedestrians or other animals of the different samples was filtered using the program Audacity 2.0.1. All the files were normalized and saved at 16 bits audio format, which is the maximum quality that this software permits. These previous digital treatments are necessary in order not to affect the results of statistical analysis (Hopp et al., 1988). We selected two minutes as the sample size used in each individual bird. Within each file, the mean frequency spectrum was calculated with the function meanspec from Seewave program in R-environment (Sueur, 2008a). The different spectra were averaged for the two sampling areas and differences between both determined by the Kolmogorov-Smirnov test (Bagdonavicius et al., 2011).

Formants, defined as the spectral peaks of the sound spectrum of the song, were determined using a forward cross-correlation [to pitch (cc) command] in Praat software (Boersma and Weenink 2013). The number of formants was determined by visually observing the bands of accumulation of energy in the spectrograms of Nightingales. Once set to 7 the minimum number of formants, the mean frequency of these was used to determine the acoustic differences between wild and urban Nightingales using a logistic regression model (Geyer 2003).

A total of 34 sound parameters were determined for each bird (Table 1). The Acoustic Complexity Index (ACI) is based on the fact that many natural sounds present an intrinsic variability of intensities compared to artificial sounds (Pieretti, et al. 2011). This index weight both the temporal variability along frequencies as the frequency variability along time giving a global measure. In the analysis of the Acoustic Diversity Index (ADI) the spectrogram is divided into bins, and the proportion of the signals in each bin above a threshold value is determined by the Shannon index (Villanueva-Rivera *et al.* 2011). Although, the ADI parameter is mainly used in soundscape analysis with various species that sing simultaneously, we consider that Nightingale calls are complex enough to justify its use. The Acoustic Evenness Index (AEI) determined the band evenness using the Gini index applied to these bins. The Bioacoustic Index (BI) is the total area formed by all frequency bands where sound surpass the minimum dB value (Boelman et al., 2007). The energy of each sound file was determined using Praat software and is expressed in Pascals2 s-1 (Table 1). The number of local peaks (NLP) in each spectrum was determined using Audacity. The parameter acoustic total entropy (H) is the product of both temporal (TH) and spectral entropy (SH) and oscillated between 0 (pure tone) to 1 (random noise) (Sueur et al., 2008b). H can be determined using seewave or soundecology packages in R (Villanueva-Rivera et al., 2011). The function specprop inside the seewave package permits determination of other parameters based on a frequency spectrum such as the inter-quartile range of frequencies (IQR), skewness (SKEW), kurtosis (KURT), mean (MEA), its standard deviation (SDM) and standard error (SEM), median of frequencies (MED), mode

(MOD), first (Q25) and third quartiles (Q75), spectral flatness (SFM), frequency precision (PREC) and centroid (CENT). Centroid is the summation along the interval of frequencies of the product between each frequency and its relative amplitude (Sueur et al., 2008a). Finally we used Audacity to divide each Nightingale's call in three parts: silent periods, rapid broadband trills and normal songs. These sequences were independently determined due to these having different behavioral functions (Schmidt et al., 2008; Sprau et al., 2010). These three sequences were used for determining the total songs per minute (S); the ratio between songs and silent periods (RSS); the percentage of the number and time of total songs (SG%N and SG%T), silent periods (SL%N and SL%T) and trills (TR%N and TR%T) and the number of songs (SGNT), silent periods (SLNT) and trills (TRNT) per minute. The 34 sound parameters of wild and urban Nightingales were compared by the Bartlett and Wilcoxon rank sum test (Bagdonavicius et al., 2011). The Bartlett test permitted determination if both populations show equal variances (homoscedasticity). Wilcoxon test is for detecting equality in the means. The combined use of both tests allow for the classification of sound parameters in four groups according to their intensity of response to urban conditions (Tables 2 and 3). In the first group the parameters without any response for both tests are included. The second group shows a light effect in variance. The third group is formed by parameters with a significant difference in the means, and finally the fourth group is constituted by the sound parameters that vary both in variance as in means. The simultaneous use of Bartlett and Wilcoxon tests had been used in different contexts of Ecology giving a precise perspective of the intensity of change in different environmental parameters (Martínez, 2009).

3. Results

The means of the normalized spectra for both groups of birds, show clearly where are the significant differences. In urban environments, nightingales increase the volume of calls under 1000 Hz. In contrast, between 1000 and 6000 Hz, the songs present less intensity. The Kolmogorov-Smirnov test shows significant differences between the spectra of both groups of birds ($D=0.201$, $p < 0.001$).

Table 2. Averages and standard deviations of 34 sound parameters determined in the songs of Urban and Wild Nightingales

Parameter	Urban	Wild	Bartlett	Wilcoxon
ACI	198.07 ∓ 21.79	208.15 ∓ 32.37	6.65 **	1340.00 ns
ADI	1.82 ∓ 0.27	1.91 ∓ 0.22	1.55 ns	824.50 ns
AEI	0.50 ∓ 0.11	0.44 ∓ 0.12	0.32 ns	1325.50 *
BI	27.66 ∓ 29.43	21.67 ∓ 26.97	0.34 ns	1259.00 ns
CENT	3845.59 ∓ 408.11	4063.18 ∓ 873.77	22.22 ***	1280.00 ns
dB	69.30 ∓ 3.80	70.20 ∓ 2.96	2.89 ns	1361.50 ns
E	0.23 ∓ 0.22	0.28 ∓ 0.25	0.95 ns	1359.00 ns
H	0.68 ∓ 0.10	0.72 ∓ 0.08	1.92 ns	1489.00 *
IQR	1833.10 ∓ 681.18	2179.43 ∓ 1372.92	19.17 ***	1303.00 ns
KURT	136.93 ∓ 570.96	145.51 ∓ 534.14	0.21 ns	1214.00 ns
MEA	3845.59 ∓ 408.11	4063.18 ∓ 873.77	22.22 ***	1280.00 ns
MED	3428.85 ∓ 359.58	3607.62 ∓ 972.19	35.27 ***	1197.50 ns
MOD	2697.82 ∓ 1041.69	3093.93 ∓ 1731.55	10.65 **	1348.00 ns
NLP	7.28 ∓ 2.03	7.16 ∓ 2.58	2.55 ns	1114.50 ns
PREC	0.04 ∓ 0.07	0.05 ∓ 0.08	2.16 ns	1369.50 ns
Q25	2676.72 ∓ 296.90	2697.93 ∓ 445.23	6.96 **	1192.50 ns
Q75	4510.59 ∓ 688.62	4877.36 ∓ 1504.10	23.26 ***	1254.50 ns
RSS	0.53 ∓ 0.86	0.29 ∓ 0.19	94.67 ***	1259.00 ns
S	10.72 ∓ 6.46	12.41 ∓ 9.95	7.85 **	1247.00 ns
SDM	2025.54 ∓ 586.14	2183.20 ∓ 711.05	1.66 ns	1321.50 ns
SEM	2.31 ∓ 0.98	2.77 ∓ 1.93	18.10 ***	1328.00 ns

Table 2. (Continued)

Parameter	Urban	Wild	Bartlett	Wilcoxon
SFM	0.06 ∓ 0.10	0.08 ∓ 0.13	6.47 *	1272.00 ns
SG%N	29.38 ∓ 7.16	27.53 ∓ 8.42	1.18 ns	930.00 ns
SG%T	46.47 ∓ 24.89	44.63 ∓ 19.54	2.77 ns	1217.50 ns
SGNT	12.60 ∓ 9.63	28.13 ∓ 56.34	87.04 ***	1444.50 ns
SH	0.88 ∓ 0.05	0.89 ∓ 0.04	1.00 ns	1324.00 ns
SKEW	5.74 ∓ 4.35	6.39 ∓ 7.19	10.39 **	1182.50 ns
SL%N	23.56 ∓ 6.71	23.68 ∓ 5.64	1.42 ns	1179.50 ns
SL%T	47.00 ∓ 24.65	43.65 ∓ 17.37	5.83 *	1019.00 ns
SLNT	10.27 ∓ 8.34	25.50 ∓ 53.93	94.21 ***	1612.50 **
TH	0.94 ∓ 0.00	0.94 ∓ 0.02	50.43 ***	1060.50 ns
TR%N	0.12 ∓ 0.13	0.15 ∓ 0.12	0.02 ns	1370.00 ns
TR%T	6.53 ∓ 6.66	11.72 ∓ 11.81	13.31 ***	1492.50 *
TRNT	0.04 ∓ 0.04	0.07 ∓ 0.07	7.62 **	1522.50 *

Bartlett test of homogeneity of variances and Wilcoxon signed-rank test for comparison of means are determined. *: $p\text{-value} < 0.05$; **: $p\text{-value} < 0.01$; ***: $p\text{-value} < 0.001$; ns: non-significant differences.

Table 3. Division of the 34 acoustic parameters based on their response to urban conditions

Parameter	Group 1	Group 2	Group 3	Group 4
ACI	-	+	-	-
ADI	+	-	-	-
AEI	-	-	+	-
BI	+	-	-	-
CENT	-	+	-	-
dB	+	-	-	-
E	+	-	-	-
H	-	-	+	-
IQR	-	+	-	-
KURT	+	-	-	-
MEA	-	+	-	-
MED	-	+	-	-
MOD	-	+	-	-
NLP	+	-	-	-
PREC	+	-	-	-
Q25	-	+	-	-
Q75	-	+	-	-
RSS	-	+	-	-
S	-	+	-	-
SDM	+	-	-	-
SEM	-	+	-	-
SFM	-	+	-	-
SG%N	+	-	-	-

Table 3. (Continued)

Parameter	Group 1	Group 2	Group 3	Group 4
SG%T	+	-	-	-
SGNT	-	+	-	-
SH	+	-	-	-
SKEW	-	+	-	-
SL%N	+	-	-	-
SL%T	-	+	-	-
SLNT	-	-	-	+
TH	-	+	-	-
TR%N	+	-	-	-
TR%T	-	-	-	+
TRNT	-	-	-	+

Group 1: no significant effects in variability (Bartlett test) and mean (Wilcoxon test). Group 2: only significant effects in variability. Group 3: significant effects only in the mean value. Group 4: Significant effects both in variability as in the mean values.

The logistic regression does not allow separation of urban and wild Nightingales by its formant structure, due to the residual deviation tested by a Chi-squared test, which was very significant ($p<0.01$). Also, none of the seven formants, presented in all the individuals (some birds showed up to nine formants), showed significant effects. Consequently, both groups of birds have similar formant structure.

Regarding the effect of urban conditions on the 34 sound parameters analyzed (Tables 1 and 2) there are significant differences between them. In the first group the parameters do not show significant effects in variability (Bartlett test) or mean value (Wilcoxon test) (Table 2). Null effects are observed in some bioacoustics indexes (ADI and BI) and average song intensity (dB) (Table 3). It is especially relevant that the total acoustic energy (E), the frequencies with more sound energy (formants) or number of local peaks in the spectra (NLP) remain unchanged under urban conditions. Also, however, there are marked differences in which frequencies are more used (Figure 2). Also, there are not any changes in the kurtosis (KURT), the frequency precision (PREC), the standard deviation of the mean frequency (SDM) or the spectral entropy (SH). Newly, the fact that these parameters are not changed show that the global sound characteristics of the calls are similar. However, some structural parameters, such as the percentage of normal songs in time (SG% T) and number (SG% N) remain unchanged too. Neither the percentage of the number of rapid broadband trills (TR% N) nor silences (SL%N) are modified under urban conditions. Therefore, nightingales use the same percentage in the total number of acoustic structures (silences, rapid broadband trills and normal songs) under urban environmentconditions. The second group of parameters, show a light response to urban conditions that only affects variability (Bartlett test) and not means (Wilcoxon test) (Table 3). This is the largest group and includes the acoustic complexity index (ACI). ACI is higher under wild conditions, although this effect of the means is not significant (Table 2). In group 2 (Table 3) of light response to urban conditions, parameters that describe global properties of the sound such as the centroid of frequencies (CENT), the inter-quartile range (IQR), the mean frequency (MEA), the median (MED), the mode (MOD) or quartiles (Q25, Q75) are presentlocated. The ratio between songs and silent periods (RSS) is apparently very different between urban (0.53) and wild (0.29) nightingales, but the differences in means are not significant (Table 2). Significant responses were also observed in the total variability of calls per minute (S) and within these for normal songs (SGNT). The spectral flatness (SFM) is another parameter that is modified in variability. Group 3 is formed by the parameters with a significant effect in means but not in variability whereas group 4 presents significance both in mean as in variability (Table 3). The acoustic evenness index (AEI) that shows how equal a sound is along the sampling time, presents differences between urban and wild conditions. AEI is based on Gini coefficient that is a measure of dispersion or inequality (Villanueva-Rivera et al., 2011). In urban conditions AEI is higher, indicating lessmore differences along time in the sound emissions. This means that wild nightingales do not have as regular calls compared to urban conespecific. This effect is consistent with the total entropy parameter (H) that is significantly higher in nightingale calls in natural environments (Table 3). Finally, the parameters that show a higher effect (group 4) are structural and analyze the changes in the types of calls during sound emissions. The number of silences per minute (SLNT) is more than half those in urban nightingales (Table 2). These urban nightingales decrease the percentage of trills in both time (TR% T) and number per minute (TRNT).

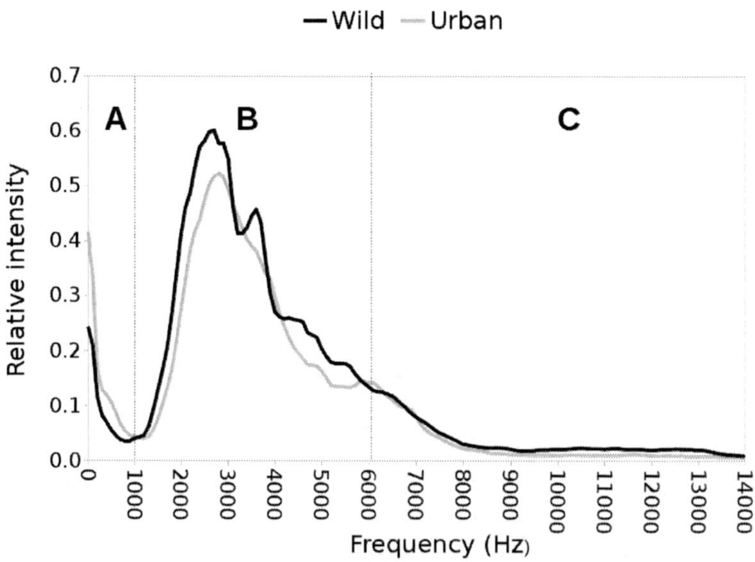

Figure 2. Normalized mean sound spectrums for Wild and Urban Nightingales. Three zones are clearly defined: A: Under 1000 Hz where Urban Nightingales emit to higher intensity. B: Between 1000 and 6000 Hz where Urban Nightingales emit to lower intensities and C: Over 6000 Hz with similar intensities in both groups. The statistical differences between both groups of spectrums were significant according Kolmogorov-Smirnov test (see text).

4. Discussion

Recent research has shown that urban noise produces a marked increase in the pitch and volume of Nightingale songs (Brumm 2004). This process can also be observed in other small urban birds (Slabbekoorn & Peet 2003). However, this study gives more detailed analysis of this sonometric information, covering a high number of individuals. This sampling effort has been necessary to determine which bioacoustic parameters are really relevant in the adaptation of Nightingales to urban areas. After determining the most significant parameters it is possible to develop new studies involving a more complex sampling. One possibility is to analyze spatial gradients that include transitional conditions between the two extreme situations analyzed in this paper, for example in exurban and suburban areas. Another option is to compare sound records along temporal gradients for determining possible evolutionary changes. In this sense, recent research has shown that some urban birds are genetically different from their wild conspecifics (Zhang et al., 2013). Nightingales, and other small songbirds, learn to sing by imitating other males (Hultsch et al., 1999). Urban noise masks the low frequencies of singing males, making these less audible. Two strategies are possible here: increase the intensity of high frequencies or reduce the volume of low frequencies (Ortega, 2012). Nightingales completely change all the acoustic spectra showing a high adaptability to urban noise. This process had been observed in many species (Brumm 2003). In the long term, it is expected that this process differentially affects reproduction and produces high speedfast evolutionary changes due to reduced gene flow of green areas in cities, very isolated from each other (Zhang et al., 2013). The observed changes between wild and urban Nightingales do not affect parameters such as the mean frequency, centroid, kurtosis, number of local peaks or formant structure. Nevertheless, it should be emphasized that the changes in the spectral emissions conserve the total energy and preserve the formant structure which is directly related to the anatomical features of the syringe (Hopp et al., 1998). Also, these changes are produced with a significant decrease in the acoustic variability of the call, defined by the total entropy (H), and an increase in the acoustic evenness (AEI). Therefore, urban nightingales present less sound variability between males and a higher uniformity along time in the sound emissions. Also, these birds use less territorial calls and are forced to emit their songs more continuously. It is possible to conclude that in urban areas the territorial pressure between males has diminished and they have less variability in their calls. Consequently, the sexual selection in urban environments must be lower compared to natural areas. This can explain why nightingale calls present the same energy in both environments. Rapid broadband trills are more energetic, but its number decrease in urban birds. In other words, the reduction in the high energy ofr rapid broadband trills is compensated by the increase in frequency of normal songs. Definitely, urban environment is attractive to nightingales and other small birds by the low presence of predators and competitors. Nightingales are migratory species that return to their birthplace each year. Consequently, the special conditions of urban environments reduce the signal to noise ratio of animal communication and have effects on a low sexual selection among males. These problems are accentuated by the isolation between bird populations in urban areas that will produce strong evolutionary changes in the future.

ACKNOWLEDGMENTS

We want to thank to several courses of students of Environmental Sciences worked selflessly in some of the censuses of the study. María Auxiladora Villegas and Florentino de Lope from Department of Zoology of UEX give useful comments in the initial version of this paper. We also thank Alastair Roy for his English corrections on this manuscript. Part of this study was financed with funds for practical classes at the University of Extremadura (SW Spain).

REFERENCES

Bagdonavicius, V., Kruopis, J. and Nikulin, M.S. (2011). *Non-parametric tests for complete data*, Iste&Wiley, London&Hoboken, 308 pp.

Boelman N.T., Asner G.P., Hart P.J. and Martin R.E. (2007). Multi-trophic invasion resistance in Hawaii: bioacoustics, field surveys, and airborne remote sensing. *Ecological Applications* 17, 2137-2144.

Boersma, P. and Weenink, D. (2013). *Praat: doing phonetics by computer* [Computer program]. Version 5.3.51, retrieved 2 June 2013 from http://www.praat.org/

Brumm, H. (2003). Animal communication: city birds have changed their tune. *Current Biology* 16(23), 2003-2004.

Brumm, H. (2004). The impact of environmental noise on song amplitude in a territorial bird. *Journal of Animal Ecology* 73, 434-440

Fernández-Juricic, E.; Sanz, R. and Sallent, A. (2001). Frequency-dependent predation by birds at edges and interiors of woodland. *Biological Journal of the Linnean Society* 73, 43-49.

Geyer, C.J. (2003). *Generalized linear models in R.* GNU Free Software Foundation. US, 23 pp.

Hopp, S. L.; Owren, M. J. and Evans, C. S. (Eds) (1998). *Animal acoustic communication.* Springer, Berlin, Heidelberg, 421 pp.

Hultsch, H.; Schleuss, F.; and Todt, D. (1999). Auditory-visual stimulus pairing enhances perceptual learning in a song bird. *Animal Behaviour* 58, 143-149.

Husté, A., Selmi, S. and Boulinier, T. (2006). Bird communities in suburban patches near Paris: Determinants of local richness in a highly fragmented landscape. *Ecoscience* 13(2), 249-257.

Jokimäki, J. and Huhta, E. (2000). Artificial nest predation and abudance of birds along an urban gradient. *The Condor* 102, 838-847.

Luniak, M. (2004). *Synurbanization. Adaptation of animal wildlife to urban development.* Proceedings 4th International Urban Wildlife Symposium (Shaw, W.W., Harris, L.K., Vandruff, L. Eds.), 50-55 pp. Tucson, Arizona.

Maklakov, A.A., Immler, S., González-Boyer, A., Rönn, J., Kolm, N. (2011). Brains and the city: big-brained passerine birds succeed in urban environments. *Biological Letters* 7, 730-732.

Martínez, M. (2009). *R for biologists.* GNU Free Software Foundation. US, 35 pp.

Naguib, M., Todt, D. (1997). Effects of dyadic vocal interactions on other conespecific receivers in nightingales. *Animal Behaviour* 54, 1535-1543.

Ortega, C.P. (2012). Effects of noise pollution on birds: A brief review of our knowledge. *Ornithological Monographs* 74, 6-22.

Patón, D. (2013). Environmental Factors that Affect Urban Avian Communities. In: *Birds: Evolution and Behavior, Breeding Strategies, Migration and Spread of Disease* (Ed. Lucas Ruiz y Franco Iglesias), 143-156 pp. Nova Publishers, N.Y. (USA).

Patón, D., Romero, F., Cuenca, J., and Escudero, J. C. (2012). Tolerance to noise in 91 bird species from 27 urban gardens of Iberian Peninsula. *Landscape and Urban Planning* 104, 1-8.

Pieretti, N., Farina, A. and Morri, D.. (2011). A new methodology to infer the singing activity of an avian community: The Acoustic Complexity Index (ACI). *Ecological Indicators* 11, 868-873.

Schmidt, R., Kunc H.P., Amrhein, V. and Naguib, M. (2008). Aggressive responses to broadband trills are related to subsequent pairing success in nightingales. *Behavioural Ecology* 19, 635-641.

Slabbekoorn, H.; Peet, M. (2003). Birds sing at higher pitch in urban noise. *Nature* 424, 267.

Sprau, P., Schmidt, R., Amrhein, V., Roth, T. and Naguib, M. (2010). Effects of rapid broadband trills on responses to song overlapping in nightingales. *Ethology* 116(4), 300-308.

Sprau, P., Roth, T., Amrhein, V., Naguib, M. (2012). Distance-dependent responses by eavesdroppers on neighbour-stranger interactions in nightingales. *Animal Behaviour* 83: 961-968.

Sueur, J., Aubin, T., Simonis, C. (2008a). Seewave: a free modular tool for sound analysis and synthesis. *Bioacoustics* 18, 213–226.

Sueur, J., Pavoine, S., Hamerlynck, O. and Duvail, S. (2008b). Rapid acoustic survey for biodiversity appraisal. *PLoS ONE* 3(12), 1-9.

United Nations (2011). Population division, population estimates and projections section. *World urbanization prospects 2011 revision* (http://esa.un.org/unup/).

Villanueva-Rivera, L. J., Pijanowski, B.C., Doucette, J. and Pekin, B. (2011). A primer of acoustic analysis for landscape ecologists. *Landscape Ecology* 26, 1233-1246.

Zhang S., Suo, M., Liu, S. and Liang, W. (2013) Do Major Roads Reduce Gene Flow in Urban Bird Populations?. *PLoS ONE* 8(10), 1-7.

In: Noise Pollution
Editor: Eugene Roberto Nicchi

ISBN: 978-1-63321-110-0

Chapter 2

DRIVER STYLE INFLUENCE IN THE VEHICLE ACOUSTIC EMISSIONS IN URBAN TRAFFIC

J. A. Calvo[1*], C. Álvarez-Caldas[1] and J. L. San Román
[1]Departamento de Ingeniería Mecánica, Universidad Carlos III de Madrid, Leganés, Spain

ABSTRACT

The attenuation of traffic noise in highly populated urban environments is one of the biggest technological challenges that the scientific community has to face. Although the legislation has gradually imposed more severe limitations to the levels emitted by vehicles, these restrictions have not been sufficient to create a sustainable acoustic environment due to the constant increase in the car fleet and because the approval test (Directive 70/157/EEC) does not represent properly the noise level emitted by a car in real driving conditions.

The European Commission has placed this issue within the priority research strategies. Following these guidelines and recommendations, this work presents a new strategy for the control of car noise emissions, and at the same time improves the understanding of the fundamental parameters that influence in the car noise emissions.

Nowadays there are no studies about the influence on global acoustic contamination due to a vehicle as individual source neither about the influence of driving style. Because of this, it is impossible to identify

* Corresponding author: jacalvo@ing.uc3m.es; Phone number: +34 91 624 8791

which drivers are responsible for generating the maximum noise levels within the traffic stream.

Results show that there is a very clear correlation between these highest levels of noise and the nuisance generated, so concrete actions for the control of traffic noise should be directed towards noisier vehicles, rather than imposing limitations on the levels produced by the total traffic stream. So, it is necessary to develop studies to correlate radiated noise and driving behaviour.

In this work, a sample of vehicles has been instrumented to obtain physical measurements of the variables that can influence on the noise emission level. Vehicle positions, vehicle speed, longitudinal accelerations and drive time have been analyzed using a GPS sensor. Engine speed, engine load, throttle position and gear shift position have been studied through the vehicle CAN BUS and a battery of microphones has measured emitted noise in several points of the vehicle.

A circuit that represents the usual circulation in a big city like Madrid has been defined, as well as several vehicle models with great presence in existing fleet of cars. Several drivers have covered the circuit at different times of the day and previous parameters have been measured in each test in order to determine average values of behavior.

To conclude, previous data have been analyzed and noise equivalent levels have been identified with different test configurations. With these average values, it has been possible to define the noise emissions produced by a vehicle in real driving conditions and to identify the noisier driving style.

As results of the investigation, it has been demonstrated that a noisy driving style produces up to 9 dB (A) more on average than other drivers. Depending on the engine type a vehicle can emit about 5 dB more of engine noise at same drive style. There is no appreciable influence of the driver experience over the considered driving interest parameters. There is no influence of the driver sex over the considered driving interest parameters.

Keywords- vehicle noise emissions, passenger car, urban traffic

1. INTRODUCTION

Traffic noise is one of the main sources of noise pollution in cities. It has become a cause of social tension between economic development and quality of life. One of the current technological challenges is to achieve a sustainable acoustic environment in constantly growing cities with a rising quantity of vehicles.

Despite the efforts made to reduce the noise emissions of individual vehicles, noise disturbance has not diminished in equal measure because it is been offset, among other things, by the growth of the vehicle fleet (1)

Current legislation obliges to determine the noise level of a vehicle according to a test procedure shown in Directive 70/157/EEC (2). If a vehicle exceeds the maximum noise level established in this procedure, it cannot be approved. This maximum noise level, which depends on the category and type of the vehicle, has been declining gradually with the various updates of the Directive. However, this reduction has not affected the overall traffic noise. One of the reasons for this fact is that the approval test does not represent properly the real noise level emitted by a real car in real driving conditions since it only considers very specific operating conditions (transient acceleration - deceleration)

To determine the noise emission level of a vehicle in a way that reproduces its work in real driving conditions, it is necessary to establish the main parameters that may affect such noise emission.

This goal is very ambitious when one considers that the type of driver, his driving style, the type of traffic, the size and type of vehicle, road conditions, etc., are factors that can influence decisively on vehicle noise emissions.

First of all, it is necessary to establish the parameters that could have more influence on noise emission. After that, a representative sample of vehicles and drivers must be selected in order to extract statistical values that represent average driver behavior. Finally, a driving route that represents an average driving condition of urban traffic has been selected.

Once all these factors have been established, several driving test have been accomplished in order to obtain statistical values that allows to determine factors that influence on vehicle noise emissions. All drivers have been committed to drive as like they usually do, except one of them. An expert driver has been committed to drive aggressively in order to compare his driving parameters and noise emissions with the average driver values.

This allows for identification of which driving behavior has more influence on vehicle noise emissions and which vehicle parameters could detect this drive mode.

2. Material and Methods

In this section every material used in the research will be presented: vehicles, instrumentation… The selection of drivers and driving route that have been used to identify the parameters that can influence in noise emitted by a vehicle during urban driving pattern will be explained also.

2.1. Sample Vehicles Selection

According to the scope of the project, only noise emission levels of passenger vehicles have been studied. These vehicles are usually classified into different segments according to their size, which corresponds well with their level of power and refinement and, therefore, their price.

Spanish vehicle fleet have been analyze, in order to select a sample of vehicles that represent current stock and allow to weigh their contribution to the overall noise. Based on this criterion the following vehicles have being selected:

- Two B-segment vehicles (compact cars): these vehicles are very abundant in the circulation of large cities. They are compact cars, usually with small engine, up to 73 kW power and low fuel consumption. If one user has two vehicles, the second use to belong to this segment. Analyzing car registrations in Spain for 2010, this segment corresponds to 30% of the fleet.
- Two C-segment vehicles (midsize cars): these vehicles are also abundant in urban traffic; they are widely used as family car and for commuting. They are mid-power (around 100 kW). Analyzing vehicle registrations in 2010, this segment corresponds to 29% of the fleet.
- Two D-segment vehicles: These vehicles are purchased by more affluent drivers. These powerful engine vehicles are also present in urban traffic. Analyzing vehicle registrations in 2010, this segment corresponds to 15% of the Spanish fleet.

The other segments have a much smaller impact on urban traffic, so its contribution to the noise level will be less important than the selected segments contribution.

For each segment, a diesel engine model and a gasoline engine model have been selected. It seems reasonable to analyze both engines although, due to economic reasons, in every segment 70% of the registered cars have diesel engine (and only 30% have gasoline engine).

Brands and models have been selected after analyzing the vehicle registration statistics published by ANFAC (3). Table 1 summarizes the characteristics of the selected vehicles.

2.2. Path Selection

The test path has been selected according to the following criteria:

- **Representativeness**: According to the report "Research for the Sustainable Mobility" (4), by the European Commission for Energy and Transport, over 75% of the EU population lives in urban areas and around a fifth of all miles traveled in the EU are urban trips. With these reasons it is clear why urban transport has so great importance in the overall mobility. The chosen circuit must be representative of the average route that a driver covers in his daily displacements in urban ways (commuting, doing arrangements for daily activities...) This is a potential 80% of trips in a big city and therefore have a high impact on noise pollution. The same report concludes that the average trip of a driver in urban roads is between 8 and 12 km. This data has been also taken into account for the selection of the circuit.
- **Traffic density**: Traffic levels in the selected area are elevated but without reaching a collapse that results in unrepresentative traffic speeds..
- **Noise emissions level:** The area has been chosen according to the "Strategic Noise Map of the City of Madrid (5), prepared by the City of Madrid, trying to assure that people in the chosen area suffer high noise emission levels due to traffic, as it is necessary for the analysis.

Table 1. Selected Vehicles

Vehicle	Segment	Engine Type	Engine Size	Vehicle Mass	Tire Size
Seat Ibiza	B	Diesel	1598 cm^3	1170 kg	185/60 R14
Seat Ibiza	B	Petrol	1390 cm^3	1049 kg	185/60 R15
VW Golf	C	Diesel	1968 cm^3	1314 kg	205/55 R16
Opel Astra	C	Petrol	1598 cm^3	1373 kg	205/55 R16
Audi A4	D	Diesel	1968 cm3	1475 kg	215/55 R16
Mercedes C180	D	Petrol	1796 cm^3	1485 kg	225/40 R18

Taking into account the above criteria, the test track showed in Figure 1 has been selected. The circuit was defined by the lines of higher levels of equivalent sound pressure level.

The circuit has a day equivalent continuous noise level between 70 - 75 dB(A) and a traffic density between 20000 - 40000 vehicles per day.

The circuit is 8500 m long and has 25 traffic lights and 3 roundabouts. It includes roads limited, generally, to 50 km/h, but some sections are limited to only 30 km/h. The 50% of the circuit runs through two-lane two- ways streets, and the other 50% runs through four- lanes two-ways streets.

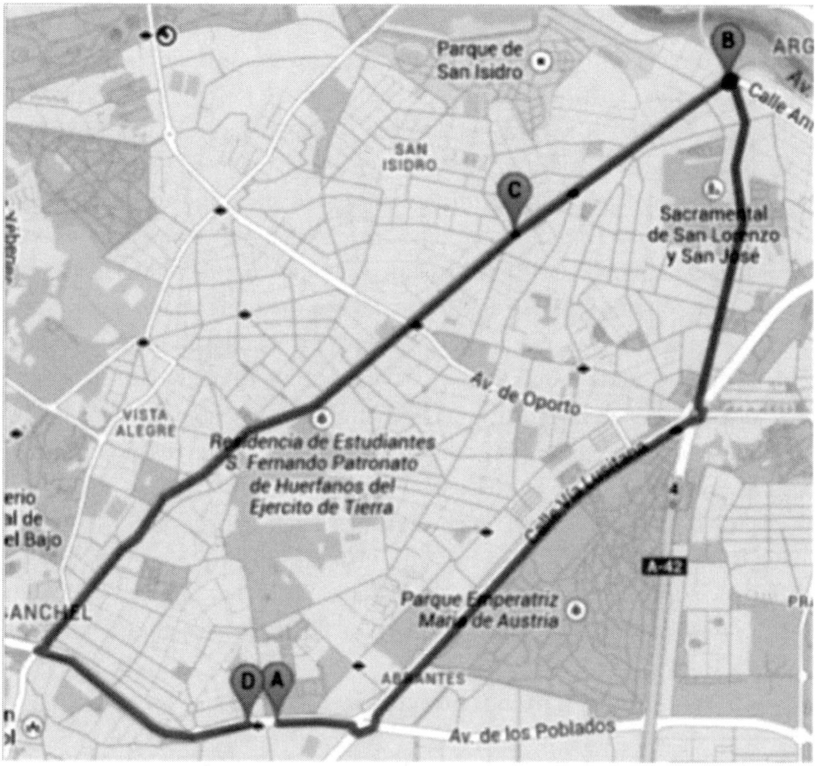

Figure 1. Selected Circuit.

2.3. Used Equipment

The following measuring equipment has been used to measure and record physical parameters in order to analyze the behavior of the vehicle:

- Acquisition and Recording System: IMC Cronos PL
- GPS Positioning System: Vbox Lite II of 10 Hz
- CAN BUS System: Module TMCAN-AOI4-Eth - ODB2. ELM317 probe
- Microphones & Conditioners: Shure MX183 Microphones. Phantom Fonestar 48 V

Figure 2 shows the used equipment as well as the position of the microphones. The first microphone is placed inside the engine room, near the entrance of the inlet valve. The second one is near the rear wheel, opposite to the exhaust pipe.

Figure 2. Details of the vehicle instrumentation.

2.4. Measured Parameters

Table 2 shows the parameters that have been recorded with the measuring devices for further analysis. Every signal is acquired and processed by the CRONOS PL and recorded in the notebook PC.

Figure 3 shows the distribution of the vehicle instrumented.

Table 2. Registered parameters

Parameter	Equipment	Units	Parameter	Equipment	Units
Engine speed	TMCAN	rpm	Longitude	VBox	Grades
Engine load	TMCAN	%	Altitude	VBox	m
Throttle position	TMCAN	%	Distance	VBox	m
Engine microphone	IMC	dB(A)	Speed	VBox	km/h
Wheel microphone	IMC	dB(A)	Longitudinal acceleration	VBox	m/s^2
Latitude	VBox	Grades	Lateral acceleration	VBox	m/s^2

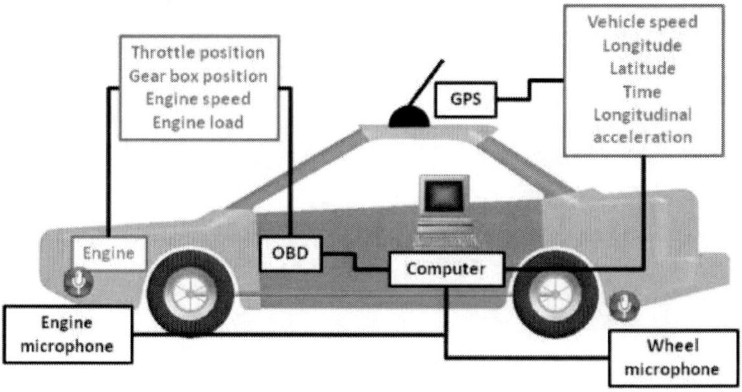

Figure 3. Details of the vehicle instrumented.

2.5. Drivers Sample

Tests have been carried out with twenty one different drivers. The sample has been selected in order to cover the different types of drivers in function of his driving skills.

- Five male, with more than 5 years of experience in driving. (Driver Category A)
- Five male, with less than 1 years of experience in driving. (Driver Category B)
- Five female, with more than 5 years of experience in driving. (Driver Category C)
- Five female, with less than 1 years of experience in driving. (Driver Category D)
- One professional driver (to conduct an aggressive driving). (Driver Category E)

This professional driver performed a test driving in an aggressive style so as to reduce travel time The aggressive style is based on high engine speeds and selected gears lower than usual in order to increase the mobility of the vehicle with continuous lane changes.

The rest of the drivers perform the circuit driving in the same way as they usually do and taking into account traffic conditions. The results of these drivers allow to compare different behaviors in order to select a representative average driver (standard driver).

2.6. Test Procedure

Once the circuit is selected and the vehicles are instrumented, the next step is taking measurements. It is necessary to arrive with the vehicle to the start point of the driving route, in order to begin the collection of data from an established reference point.

Tests have been carried out by the different drivers randomly. Drivers perform the circuit driving in the same way they usually do and taking into account traffic conditions. Tests have been carried out at several hours (morning and afternoon) and in several days of the week. The driver is always accompanied by two people who operate the measurement equipment. Once the driving route is finished, the vehicle returns to the workshop to change the driver and store acquired data in the workgroup server.

3. Results and Discussion

3.1. Driver Parameters

In this chapter obtained data are analyzed, paying special attention to those results that have helped to define the principal parameters that allows to estimate the noise emissions produced by a vehicle in real driving conditions.

The first studied parameter is vehicle speed. The vehicle speed has been recorded only when the vehicle is moving. It has not been taken into account the stopping time to evaluate vehicle speed average values.

Figure 4 shows the average of the vehicle speed in the full driving routes. Differences in the average vehicle speed between driver categories are small except in the case of the aggressive driver (driver category E). Likewise, differences between different vehicle types are insignificant.

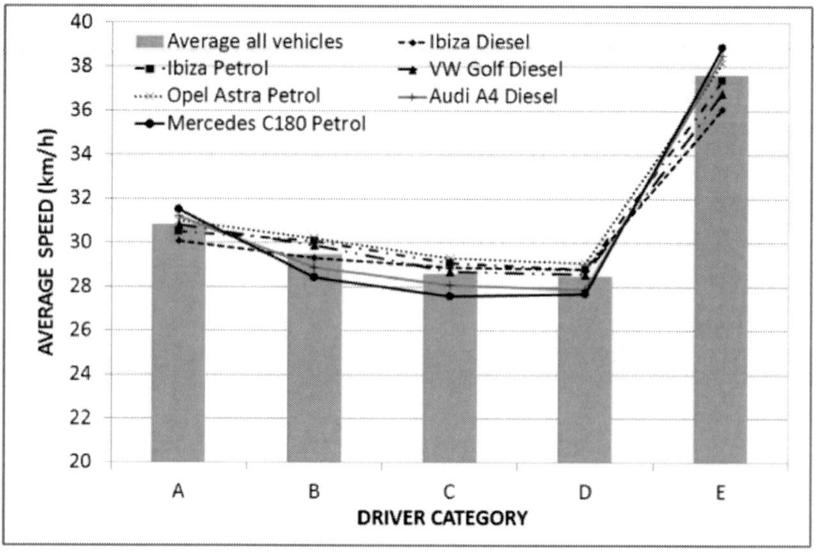

Figure 4. Average of Vehicle Speed.

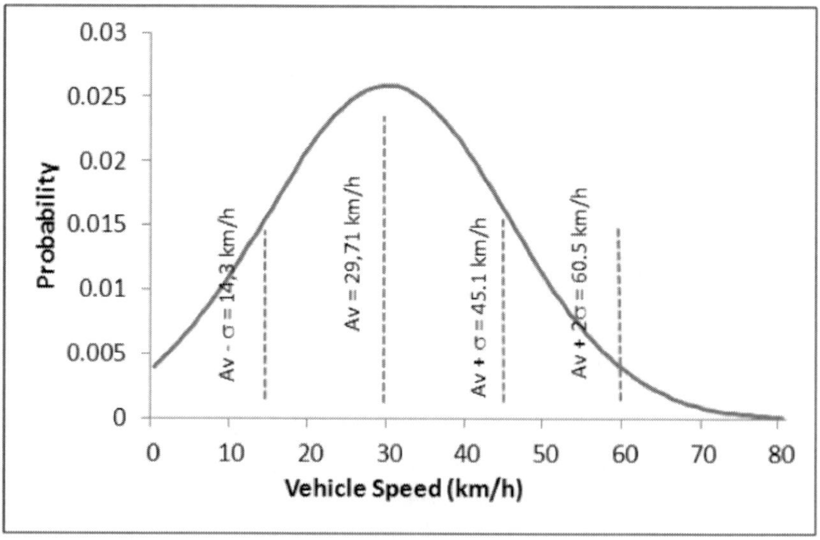

Figure 5. Probability Distribution of Vehicle Speed.

When considering all the drivers and all the vehicles it is possible to assume that the distribution is a Normal function, then, the probability distribution of the vehicle speed can be seen in Figure 5. The driver *E* has been excluded of the average calculation.

This is typical of urban traffic, where speed is limited in most sections to 50 km/h.

The average value of the standard drivers (*A+B+C+D*) is 29.7 km/h in contrast to the *E* driver that achieves an average value of 37.6 km/h (9.7 km/h higher). It is important to highlight that to achieve an average value of 29.7 km/h it is necessary that the drivers exceed the speed limit at least a 17 % of the trip time.

The average of the maximum vehicle speed for the standard drivers (A+B+C+D) was 68 km/h and for the aggressive driver (E) was 100 km/h. This means that the aggressive driver exceeds twice the maximum allowed speed.

The novice drivers(C & D drivers) are slower with the larger vehicles. Men with more experience (A category) are significantly faster than the other categories for all vehicles. The difference in the average speed for all vehicles and all categories of drivers (*A+B+C+D*) is insignificant, especially compared to the difference between the aggressive driver and any other.

Figure 6 shows the standard deviation of the vehicle speed. Standard drivers have a small value (lesser variation of the vehicle speed) that means a smooth driving. The aggressive driver has a higher variation of the speed (27 % higher from the average).

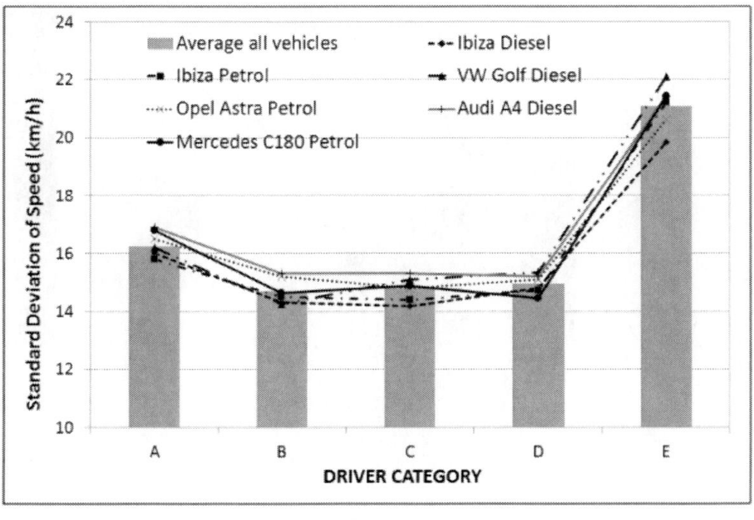

Figure 6. Standard Deviation of the Vehicle Speed.

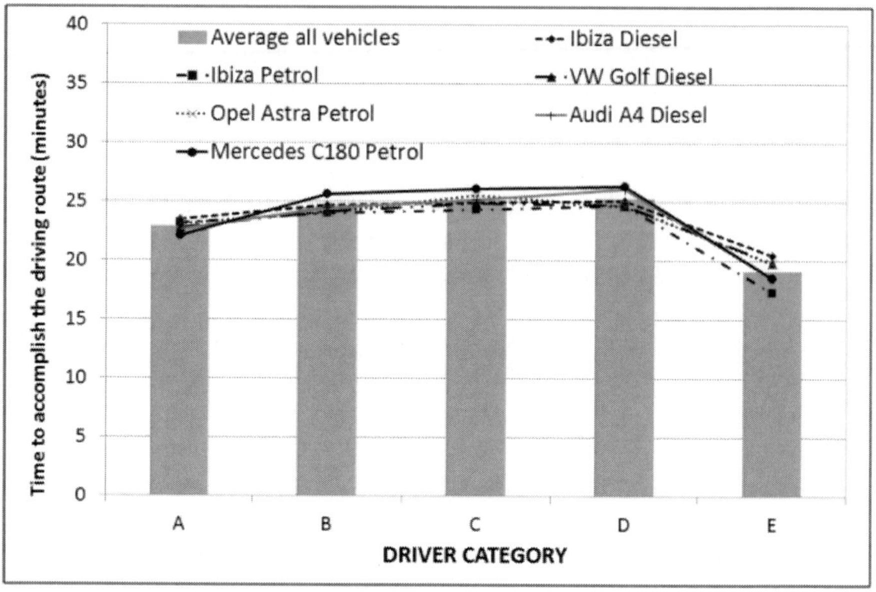

Figure 7. Time to accomplish the Driving Route.

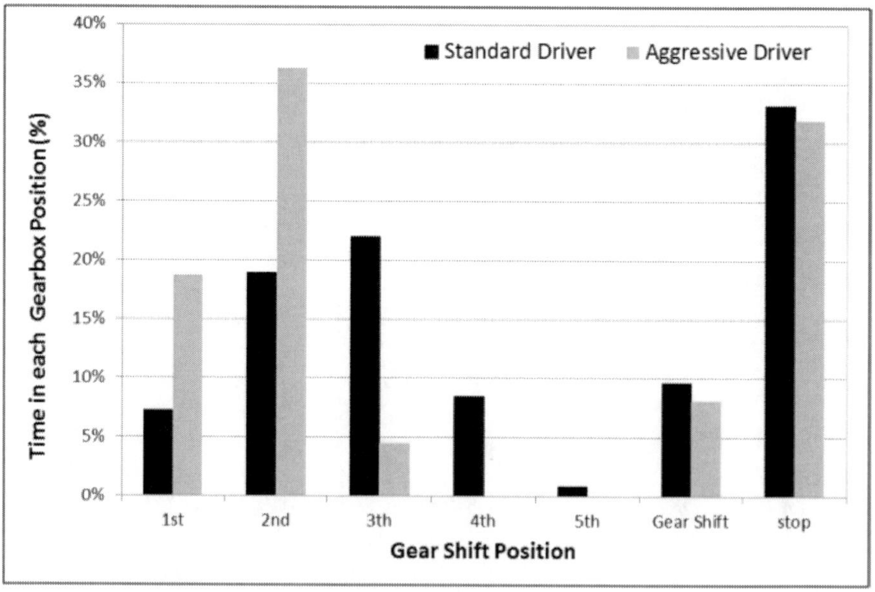

Figure 8. Time in each Gearbox Shift.

Figure 7 shows the time employed to accomplish the driving route. Differences between the standard drivers (A+B+C+D) are small and very similar for all vehicles. The aggressive driving style allows to reduce 5 minutes over the average of the rest of the drivers (20 % better).

Figure 8 shows the percentage of the time spent in each gearbox position (average of all vehicles). Standard drivers used preferably the second and third gearbox position. The aggressive driver (E) used the first and second ones.

The time during which the vehicle is stopped is very similar in all cases and it depends mainly on traffic conditions, traffic lights, etc. The time during which the driver is shifting gears is slightly smaller in the case of the aggressive driver. Both values seem independent of the driver and vehicle in case of standard drivers.

Engine speed has been analyzed in the same way. Figure 9 presents the average of all vehicles and all standard drivers versus the aggressive one. In the case of standard drivers, the most used ranges are 1000 - 1500 rpm and 1500 - 2000 rpm, which match the zone of maximum torque and therefore of minimum consumption. However, the aggressive driver used the engine at higher speeds (2500 – 3000 rpm) and the dispersion of the chart is higher.

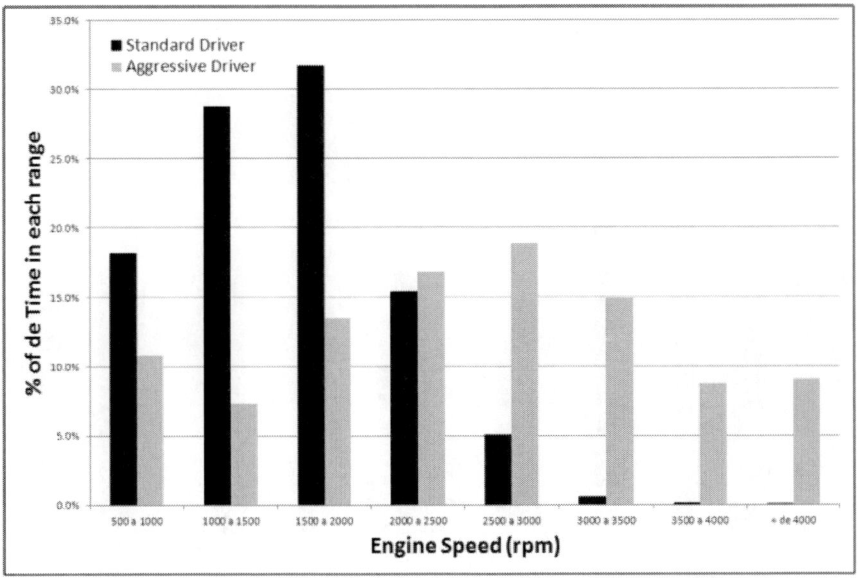

Figure 9. Time in each range of engine speed (express in %).

Figure 10 shows the normal probability distribution for both cases.

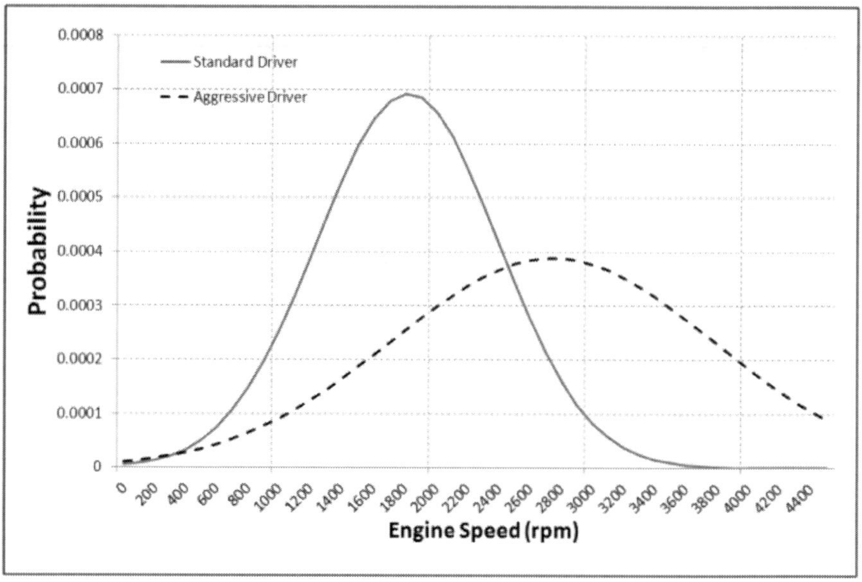

Figure 10. Probability distribution of engine speed.

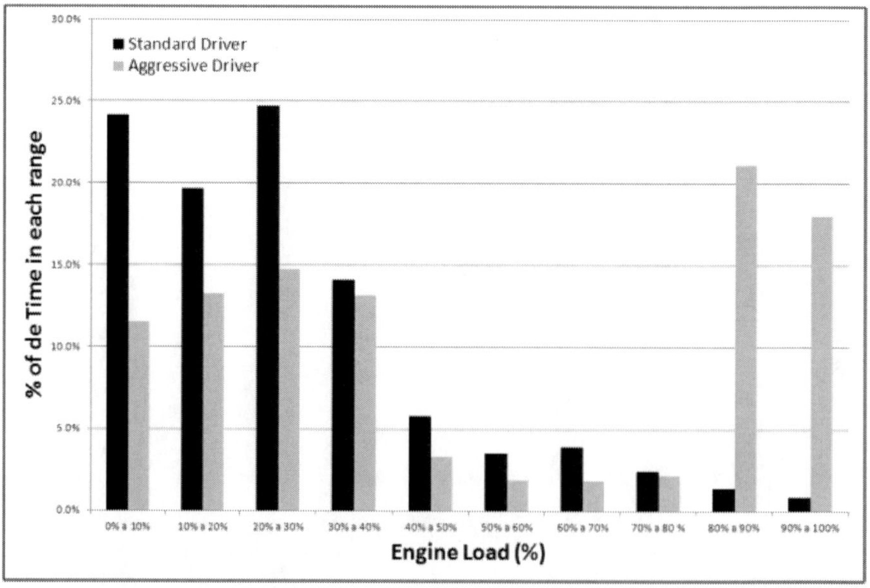

Figure 11. Time in each range of engine load (express in %).

The average value for standard drivers is 1800 rpm and the standard deviation 575 rpm. In the case of the aggressive driver, the average value is 2750 rpm and the standard deviation 1026 rpm. The standard driver leads to lower engine speeds and also less variation on the average (smooth driving). The aggressive driver leads to higher engine speeds and also has greater variation. This kind of driving increases the fuel consumption and pollutant emissions (gas and noise).

Figure 11 analyzes the time spent in each range of engine load for every car and every driver.

The average for standard driver is 30% of maximum engine power, which means that the vehicle engines are oversized to travel on urban routes.

4. Noise Emissions Analysis in Both Circuits

To analyze the influence of the considered driving variables and driving interest parameters over emitted noise $L_{eq,1\,s}$, graph from microphone 1 (engine compartment) and Levels Histogram were built for each vehicle–driver combination. Since drivers A to D appeared to show very similar results in driver parameters, noise measured data is depicted for just the average of them (referred as "standard driver") against the driver E ("aggressive driver").

Figure 12 analyzes the evolution of the $L_{eq,1s}$ of the engine microphone, for standard drivers and aggressive one. The upper chart shows the results for diesel engine and the bottom one for petrol engine. It can be observed that the sound level obviously decreases when the vehicle is stopped and increases when the vehicle is accelerating. Notice also that engine noise is higher for the diesel than for the petrol vehicle, mainly at minimum values (neutral gear, traffic lights, sense changing).

Figure 13 shows the evolution of the $L_{eq,1s}$ of the wheel microphone for standard drivers and for the aggressive one. The upper chart shows the results for diesel engine and the bottom one for petrol engine. As expected, the wheel noise is more correlated with the speed of each vehicle. Therefore, the wheel noise emitted by the aggressive driver is significantly higher than the wheel noise emitted by the standard driver. No significant differences have been found between the gasoline engine and the diesel one in the case of wheel noise.

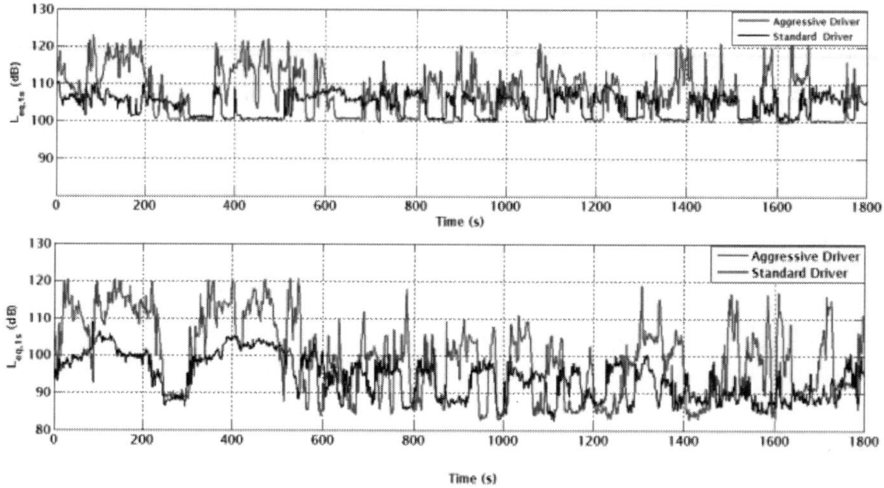

Figure 12. Results of the Leq,1s graph of the Vehicle Engine Noise.- Diesel Engine (Top) - Petrol Engine (Bottom).

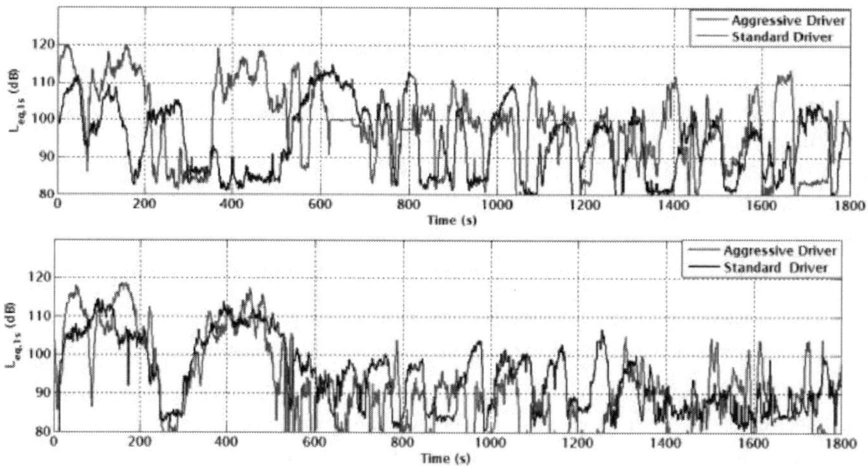

Figure 13. Results of the Leq,1s graph of the Vehicle Wheel Noise.- Diesel Engine (Top) - Petrol Engine (Bottom).

Following, Figure 14 shows the sound pressure level histograms for the engine noise produced by the aggressive driver in comparison with standard drivers. Basically the first peak in the plots represents when the vehicle is in neutral. Two charts are presented: for diesel (top) and gasoline (bottom) engines. The histogram of aggressive driver is displaced towards higher levels in both cases. As expected, this displacement is more significant for the diesel engine.

Figure 15 shows the sound pressure level histograms for the wheel noise produced by the aggressive driver in comparison with standard drivers.

The analysis of the obtained results shows:

- It is possible to find differences in the $L_{eq,1\ s}$ graph and in the Sound Pressure Levels Histogram for the aggressive driver compared to the standard one in the same vehicle, which indicates that the driving style influences the emitted noise level.
- It is very easy to associate the spectrogram shape to the vehicle characteristic noise.

Table 3 compares driver E (Aggressive) L_{eq} with standard drivers A-D L_{eq}. Aggressive driver is 9 dB noisier for the engine noise and 4.7 dB noisier for the wheel noise. In case of standard driver the wheel noise is similar to engine noise (0.7 dB of difference). However, the aggressive driver produces higher level of engine noise than wheel noise (3.6 dB).

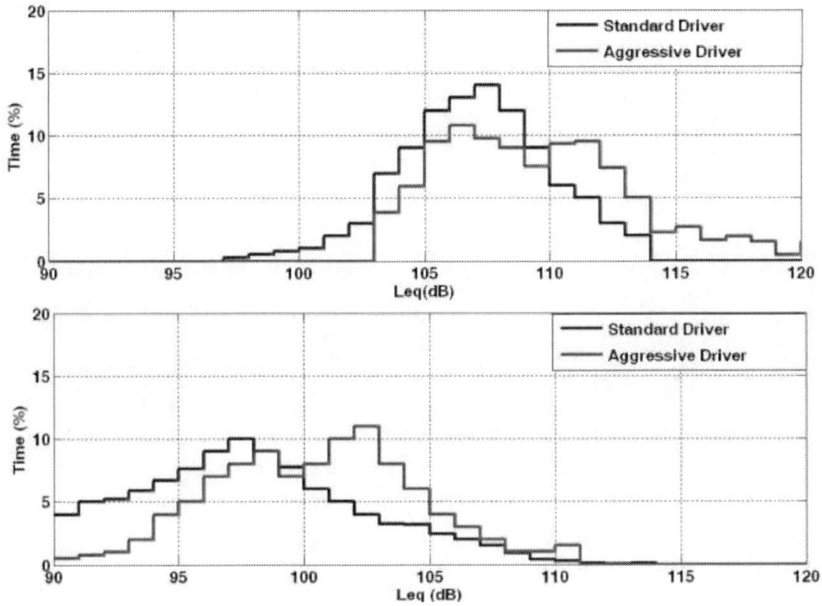

Figure 14. Sound Pressure Level histograms graphs for Engine Noise - Diesel engine (Top) – Petrol Engine (Bottom).

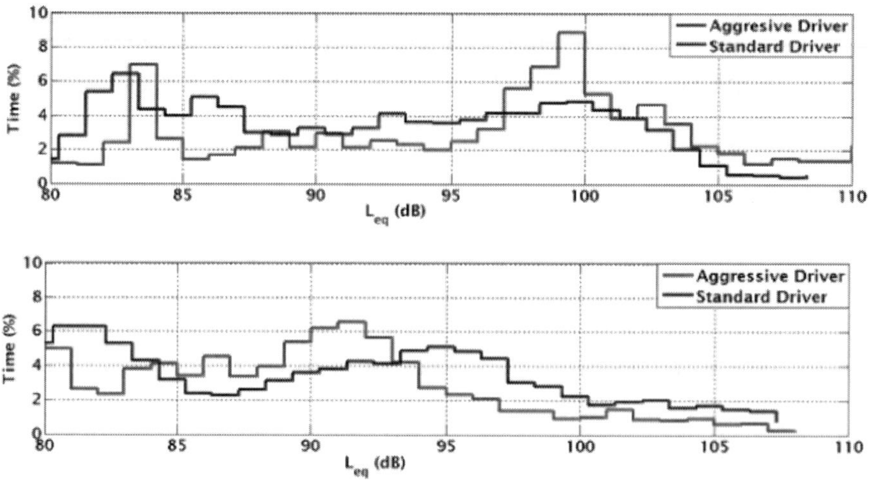

Figure 15. Sound Pressure Level histograms graphs for Wheel Noise - Diesel engine (Top) – Petrol Engine (Bottom).

Table 3. Global equivalent levels L_{eq} for engine and wheel noises for every driver - Petrol engine

Driver	Engine Noise (dB)	Wheel Noise (dB)
A	103.2	104.5
B	102.6	103.5
C	102.2	102.6
D	102.3	102.5
Average (A-D)	102.6	103.4
E	111.7	108.1

Table 4. Global equivalent levels L_{eq} for engine and wheel noises for every driver – Diesel engine

Driver	Engine Noise (dB)	Wheel Noise (dB)
A	107.9	106.8
B	106.9	106.9
C	106.8	105.9
D	106.6	105.6
Average (A-D)	107.1	106.3
E	114.5	110.4

In Table 4 it can be found again that the aggressive driver is 7.4 dB noisier for the engine noise and 4 dB noisier for the wheel noise. In case of standard driver the wheel noise is similar than engine noise (0.7 dB of difference). However, the aggressive driver produces higher level of engine noise than wheel noise (4.1 dB)

It can finally be deduced that depending on the engine type a vehicle can emit about 5 dB more of engine noise and 3 dB more of wheel noise if it is driven in a “noisy” way. However, it is important to highlight that the results of wheel noise are less representative than engine noise because of the kind of used circuit.

Wheel noise has been measured in conditions of “open circuit”, in real traffic, and it wasn’t possible to control those conditions. On the other hand, it is well known that wheel noise is mainly related to wheel speed and the speed is not very different between all drivers in an urban circuit. From this point of view, the results of engine noise are considered more representative of the driver behaviour.

Analyzing the results, it has been confirmed that global noise levels and Levels Histograms are useful to discriminate between different driving ways. And they are especially useful to identify noisy drivers.

CONCLUSION

The average behavior of a number of vehicles and of a sample of drivers has been analyzed. The circuit that has been considered is an urban circuit with high level of traffic noise.

According to results, a pattern that reproduces the average behavior of standard drivers has been established. That allows to determine the noise emitted by a vehicle in real driving conditions by a standard driver and it is possible to discriminate a noisier driver

By registering the noise emitted in this test, it will be possible to establish limits that allows to measure the acoustic impact of a vehicle. The next step of this study is to determine those limits by measuring noise emissions of a significant number of vehicles.

Comparing results in the driving parameters for the different driving variables combinations, some conclusions can be obtained:

- There is no appreciable influence of the driver experience over the considered driving interest parameters.

- There is no influence of the driver sex over the considered driving interest parameters.
- There is no high influence of the engine type (Petrol - Diesel) over the considered driving interest parameters - maybe a small influence over the engine speed, but not very significant.
- Driving condition (Aggressive - Normal) has definitely influence over engine speed and selected gear, but not significantly on vehicle average speed. Maximum speed and speed dispersion can be nevertheless influenced.
- Regarding obtained noise measurements, it can be concluded:
- The most relevant parameter to compare noise emissions between a standard driver and an aggressive driver is the engine noise.
- Wheel noise is significantly higher for aggressive drivers than for standard driving conditions. This difference is more remarkable for diesel vehicles than for petrol ones.
- Engine noise is also significantly higher for aggressive driving conditions, especially for petrol vehicles because they have a higher engine speed limit. For low and medium engine speeds, petrol vehicles produce less noise than diesel ones.
- In analogous conditions, engine noise is always higher for diesel vehicles than for petrol ones.
- For diesel vehicles, the engine noise is higher than wheel noise for any kind of driving conditions (aggressive or not). On the other hand, for petrol vehicles engine noise is almost equal or even lower than wheel noise under standard driving conditions.
- When driving under aggressive conditions engine noise is always higher than wheel noise.
- Aggressive driving conditions produce higher equivalent noise levels than standard ones.
- By performing more aggressive driving, average speed can be minimally increased and travel time can be slightly reduced. To achieve these goals, the aggressive driver has to exceed twice the maximum allowed speed. The noise emitted driving aggressively are quite higher than the levels achieved driving softer. Therefore, benefits of an aggressive driving are ridiculous compared to the inconvenience caused by it.

ACKNOWLEDGMENTS

This work has been supported by the Spanish Ministry of Science and Innovation (MICINN) through Grants Nº TRA2008-05654-C03-02 and TRA2008-05654-C03-03.

REFERENCES

[1] Affenzeller, J. and Rust, Al., 2005. Road traffic noise- a topic for today and the future, VDA Technical Congress, Ingolstadt (Germany), March, 16th- 17th.

[2] EEC, 1970. Council Directive 70/157/EEC of 6 February 1970 on the approximation of the laws of the Member States relating to the permissible sound level and the exhaust system of motor vehicles, T. C. o. t. E. Communities.

[3] ANFAC, 2009. Matriculaciones turismos y vehículos comerciales. Diciembre 2009.

[4] Akerman, J., Banister, D., Dreborg, K., et al. (eds), 2000. European Transport Policy and Sustainable Mobility.

[5] Madrid, Ayuntamiento de, 2006. Mapa Estratégico de Ruido de la Ciudad de Madrid.

In: Noise Pollution
Editor: Eugene Roberto Nicchi
ISBN: 978-1-63321-110-0

Chapter 3

Reaction to Hybrid Noise in Communication

***Sheuli Paul*[1,*], *Michael M. Richter*[1,2,†] *and Volker Michel*[3,‡]**
[1]University of Kaiserslautern, Germany
[2]University of Calgary, Canada
[3]University of Siegen, Germany

Abstract

The noise pollution is a vast subject. Noise can be controlled and much of the technology for solving most noise problems is available today. To control the noise, there are many noise measurements, monitoring devices and techniques. However, the noise pollution and its effects on the environment is not really improved. We are interested in how far noise pollution disturbs communication and what can be done. Here, we provide three different case studies where we discuss how an effective communication can place in a noisy environment. We consider three different types of noise such as mild, time varying steady-unsteady and strong. As situations we considered experimental laboratory, health care support, and diagnostic support for persons with possible dementia. In this chapter, we describe how a human and a machine can communicate in these three scenarios in order to make the workplace and resources more productive.

*E-mail address: paul@eit.uni-kl.de
†E-mail address: richter@informatik.uni-kl.de
‡E-mail address: michel@mathematik.uni-siegen.de

1. Introduction

We have provided three scenarios where mild, time varying steady unsteady and strong noise are very common. There have been many attempts for treatments but they have in common that they looked at each noise type individually. In our scenarios, they occur, however, simultaneously in an unpredictable way. We handle them in a way which we call hybrid. For this we have developed a hybrid speech enhancement model. The enhanced speech is recognized by our recognition model. The whole recognition system is embedded in the machine so that the machine is a replacement to a human being. Thus, the noisy situation is actively controlled in order to create a productive working environmement and an effective communication in spite of hybrid noise. In this chapter, we provide our solution approach for the hybrid noisy speech for its recognition.

There are three major situations we will present:

1. Experimental laboratory: Giving commands in a laboratory to a machine which automatically performs it.

2. Health care support: Persons who are sick and/or unable to move may want to give commands that are automatically executed by a machine without calling a nurse.

3. Diagnostic support for persons with possible dementia: These persons are asked questions like "What is the day of the week today?" The answer is checked automatically. If it is wrong then a medical doctor is called.

In all three situations, the speakers are humans and the receivers are machines. It is essential that the machines understand the commands correctly. In a clean environment today this is technically possible. This is not anymore the case if noise is present. The acoustical environment in a room depends greatly on the size, shape, and other properties of the confining walls, floor, and ceiling. In addition, some external noise can be present as we will explain in our situations.

The type of noises can be characterized applying source, level, extremity. Here, we narrow our focus on the noisy sound, its effect and its type of extremity. First we give a brief overview of our noise types.

We do not consider arbitrary noise but only three :

- **Mild:** This type of noise is soft but always exists in the environment. This may make the communication annoying to the humans who can hear well or who may not have sharp hearing ability. Moreover, this soft mild noise is disturbing to some extent, this type of noise has still to be treated for a noise robust speech recognition system.

- **Steady-unsteady time varying (Time varying):** This noise is continuous and ongoing in a steady manner or with a little variation with time. This type of noise is treated by applying adaptive filtering. In such noisy environment, communication among human beings may be possible but to make the work place productive, we suggest to replace the environment using the speech recognition system.

- **Strong:** This type of noise is extremely noisy such that human beings have difficulties to understand each other. We do not deal with longer periods of strong noise. Instead, we have two assumptions:

  1. The strong noise lasts for a very short time only.
  2. The strong noise is randomly distributed and does not occur very often.

  A typical example of a strong noise is a sudden burst or a loud sound of a heavy object falling down. In such case we apply an approach which identifies the noise and detects this for its deletion from the speech for the machine recognition. This deletion may delete a part of the speech that need to be understood. For this, a feedback is necessary to the speaker.

These three types of noise occur regardless of the types of the sources. Controlling the sources in general is not possible. By this, we mean we cannot keep production machines in the industry shut down in order to control the noise. This may have a negative impact on the economic condition. But we may control the noisy environment by applying speech recognition technology in the environment where machines can be replacements for the human beings.

Treating the noise enhances the communication in the noisy environment, increases work place productivity and improves health care because people are not required to expose themselves very often to the noise and a machine which has an embedded a speech recognition system, can perform the task.

In figure 1, we see an example of a mixed noisy signal in the time and the frequency domain.

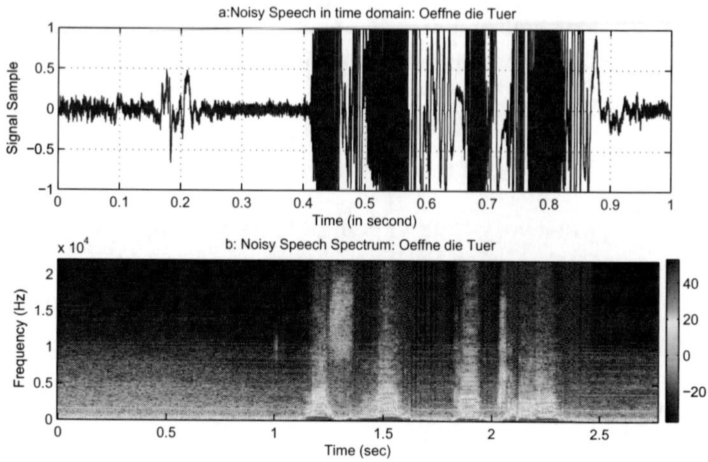

Figure 1. Hybrid noisy signal: Noisy speech signal in the time and in the frequency domain.

Figure 2 shows sound levels of the noisy signals using an A-weighting filter. The data are collected at different times from the mixed noisy environment. This figure implies the varying sound levels of the signals. We provide here the speech recognition system for the nursing home as an assistance to the elderly disabled or handicapped person. We also provide the speech recognition system for the health care to assist the nursing staff to support the mentally disabled people. All these are discussed in this chapter. These all has a basic approach and for these purposes, feature extraction and pattern matching techniques play a central role.

We will provide a model, solutions and experiments for this in the chapter.

2. First Scenario

In a medical ward, mentally disabled people are unable to communicate consciously and the nurse has to monitor them continuously. They may not be able to use distinct language in order to communicate with others. Sometimes, they can communicate using a clear voice but it may be inconsistent. They sometimes work unconsciously. They need to be monitored and observed all the time and the report has to be given to the medical expert while the serving nurse is unable to take care of them. This communication is done here by using a speech

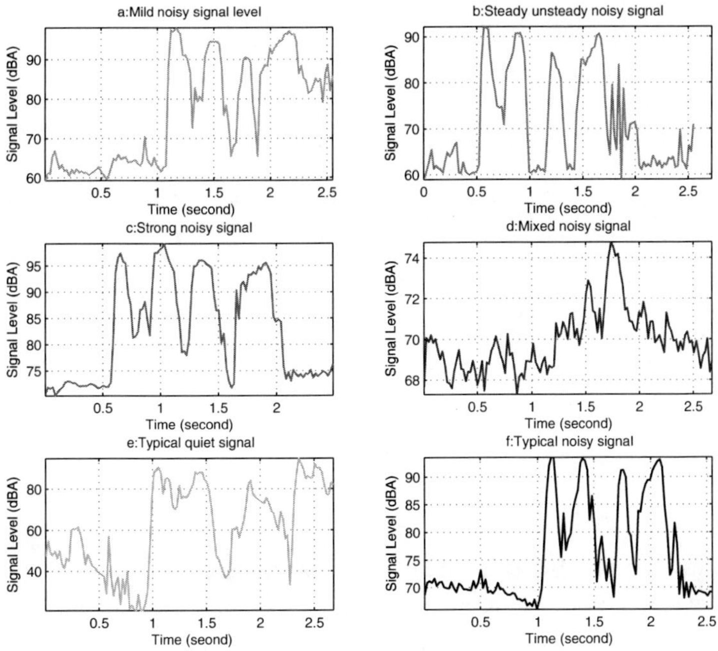

Figure 2. Hybrid noisy sound level measurements by an A-weighting filter.

recognition system. For this, we collect data from the place, extract features using the MFCC feature extraction technique, and recognize the speech using the Dynamic Time Warping (DTW) technique.

2.1. MFCC Analysis

The MFCC tries to approximate the human hearing. This extracts perceptual speech features for the speech recognition. It is probably the first perceptual speech feature extraction technique for the speech recognition system and is examined in [14]. This is one of the most commonly used feature extraction techniques in the automatic speech recognition (ASR) technology. Figure 3 gives an overview of the MFCC feature extraction technique. Next we give a short description of the components.

Pre-emphasis: First, the input signal denoted by $\acute{s}[n]$ is pre-emphasized by using a pre-emphasis filter. In equation (1), the pre-emphasized signal is $s[n]$

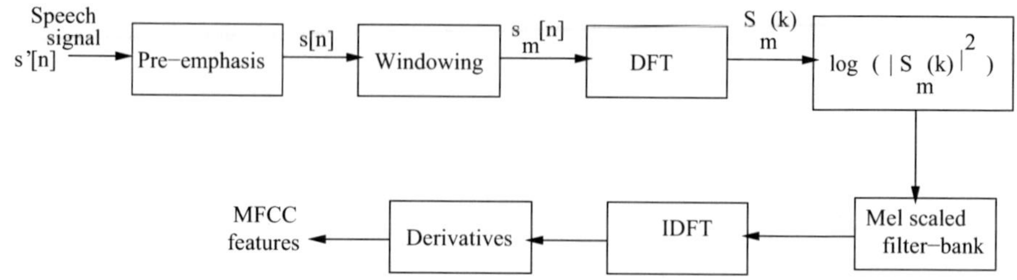

Figure 3. MFCC feature extraction technique.

for $n \in \mathcal{Z}$ (integers) and $a_{em} \approx 0.97$.

$$s[n] = s^{'}[n] - a_{em}s^{'}[n-1] \tag{1}$$

Windowing the Signal: The signal is framed into m blocks by applying a window function of length N in such a way that each signal of the m blocks has N many samples where $m = 1, 2, \cdots, M$ and $n = 0, 1, \cdots, N-1$. This segmented signal is called $s_m[n]$ by equation (2). $s_m[n]$ is also known as a frame. The framed signal or the windowed signal $s_m[n]$ is a multiplication of the signal $s[n]$ by the window function $w[n]$. Multiplying the signal $s[n]$ by the window function $w[n]$, we attain the signal frames denoted by $s_m[n]$ for $m = 1, 2, \cdots, M$ and $n = 0, 1, \cdots, N-1$, where $r \leq N$ is the shift of the samples between adjacent segments. Then $s_m[n]$ is used for the power spectrum computation in the next stage.

$$s_m[n] = s[n+mr]w[n] \tag{2}$$

Logarithmic Power Spectrum: In this step, first the logarithm of the square of the spectrum of the windowed signal $S_m(k)$ of the windowed signal $s_m[n]$ is computed by taking the logarithmic computation of the square of the absolute value of the magnitude of the DFT for the windowed signal. This is shown in equation (3).

$$S_m(k) = 10\log_{10}(|\sum_{n=0}^{N-1} s_m[n]e^{-j2\frac{\pi}{N}kn}|^2) \tag{3}$$

Perceptual Spectral Analysis: At this step, the filtering processing of the hearing and the perception of sound are attempted. The human ear has a high frequency resolution in the low frequencies and a low frequency resolution in the high frequencies. In order to reflect the frequency resolution property of the human ear, the logarithmic power spectrum shown in equation (3) is multiplied by the Mel scaled triangular filter banks. The Mel scaled filter banks given in equation (4) is known as Mel filter bank. This is a set of triangular shaped filter banks which is scaled by a Mel scale. In equation (4), the number of filters is M and $m = 1, 2, 3, \cdots, M$, where m denotes the triangular filter and f is the frequency in each bin whose size is equal to the FFT size. Now, the filter-bank $H_m(k)$ is given in equation (4). $f(m-1)$, $f(m)$, $f(m+1)$ are the left, middle and right boundary of the mth filter. $H_m(k)$ is the weight of energy at frequency k for the mth filter [6].

$$H_m(k) = \begin{cases} 0 \quad \text{for } k < f(m-1) \\ \frac{2(k-f(m-1))}{(f(m+1)-f(m-1))(f(m)-f(m-1))} & \text{for } f(m-1) \leq k \leq f(m) \\ \frac{2(f(m+1)-k)}{(f(m+1)-f(m-1))(f(m+1)-f(m))} & \text{for } f(m) \leq k \leq f(m+1) \\ 0 \quad \text{for} \quad k > f(m+1) \end{cases} \tag{4}$$

Here f_l and f_h are the lowest and highest frequencies of the filter-bank in Hz, and f_s is the sampling frequency in Hz, M is the number of filters, and N is the size of the FFT. Then the boundary points B are non-uniformly spaced in the mel-scale. One bin has the same length as the size of the FFT.

$$B(f) = \frac{N}{f_s} B^{-1}\left(B(f_l) + m\frac{B(f_h) - B(f_l)}{M+1}\right) \tag{5}$$

The Mel scale is given in equation (6). This scale is linearly spaced up to 1000 Hz and logarithmically spaced above 1000 Hz.

$$B(f) = 2595 \log_{10}(2 + \frac{f}{700}) = 1127 \log_e(1 + \frac{f}{700}) \tag{6}$$

The inverse expression $B(f)$ is the transformation of the Mel scale to the linear frequency scale Hz as shown in equation (7).

$$B^{-1}(f) = 700(10^{(f^{-1}(m)/2595)} - 1) = 700(e^{(f^{-1}(m)/1127)} - 1) \tag{7}$$

In equation (8), $H_l(k)$ is the Mel filter bank coefficient for $l = 0, 1, 2, \cdots, L$ where L denotes the number of filters that are mapped to the Mel scale and

$k = 0, 1, \cdots, \frac{N}{2}$. Due to the FFT symmetry property, the computation is taken for $0 \leq n \leq \frac{N}{2}$ which is one half of the total length N. In equation (8), $\aleph_m(l)$ is the perceptual power spectrum of the lth critical band in the Mel scale and the subscript m denotes the power spectrum of the mth frame. $H_l(k)$ is defined in equation (4).

$$\aleph_m(l) = \sum_{k=0}^{\frac{N}{2}} H_l(k) S_m(k) \tag{8}$$

Logarithmic Perceptual Power Spectrum: In this step, the logarithm of the squared magnitude of the output of the Mel filter bank is computed. The logarithm of the spectrum of each filter-bank is obtained on each frame using equation (9). The reason for computing the logarithm on the Mel power spectrum of the speech frames is to compress the wide-ranging varieties to follow the dynamic range compression characteristics of the human hearing system. In equation (9), the logarithmic power spectrum of each filter is $P_m(l)$ for $l = 1, 2, \cdots, L$ many triangular band pass filters and $m = 1, 2, \cdots, M$.

$$P_m(l) = 10 \log_{10} | \, [\aleph_m(l)] \, | \tag{9}$$

Feature Transformation Applying DCT: Since the log power spectrum is real and symmetric, the inverse DFT is equivalent to the discrete cosine transformation (DCT). The detailed information about this computation and the related transformation can be found in [13]. The DCT has the property to produce uncorrelated features and thus the features variations can be set using a diagonal assumption. This property is used to model the speech features using the Gaussian mixture model (GMM) for the classification in the HMM recognition. But this type of recognition is not considered here. We use the MFCC features for pattern matching using a dynamic time warping (DTW) technique.

Now, the DCT of $P_m(l)$ is computed by the formulation given in equation (10) where $l = 1, 2, \cdots, L$ and $n = 0, 1, 2, \cdots, J$. The notations () and [] are used to denote the time and frequency domain index.

$$c_m[n] = \sum_{l=1}^{L} \left(P_m(l) \cos(\pi n(l - 1/2)/L \right) \tag{10}$$

The $c_m[n]$ are the standard coefficients of the MFCC features. But more dynamic information of the speech can be captured by taking derivatives of the speech. However, we have not considered this here.

2.2. Dynamic Time Warping (DTW)

This technique originated from dealing with general temporal processes when one is searching for the nearest neighbor of a given process. For this, the DTW uses mainly the similarity measurement. In this approach, the unknown speech features are used in the known speech features to find the best match of the unknown speech features. The reference speech features are known speech features and the test speech features are unknown speech features. The DTW measures the similarity between the test and reference speech features to find the best match. The similarity (or equivalent distance) measurement allows to compare two processes. This similarity is measured in two steps:

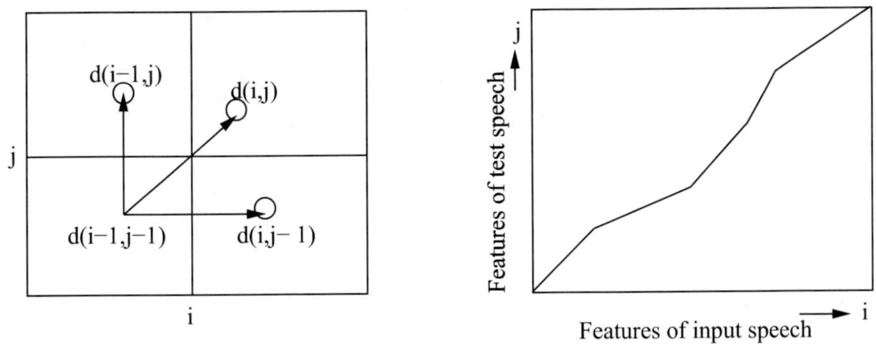

Figure 4. a: Computational Approaches of DTW and b: DTW alignment path in the speech features.

- Global similarity measure: Here, the similarity of the features between the entire reference and test speech is investigated. This formulation is shown in equation (11). Each point in the matrices is marked as node (it is shown in figure 4 by using an arrow). Each node such as i, j is associated with costs. There, the global DTW path $D(i, j)$. This is computed up to node (i, j) where the local distance at (i, j) is given by $d(i, j)$. The global similarity measure results in global DTW path as shown in figure 4, where the features of the reference speech signal are in the vertical direction and the features of the test speech signal are in the test or unknown signal is in the horizontal direction.

$$D(i,j) = min[D(i-1,j-1), D(i-1,j), D(i,j-1)] + d(i,j) \quad (11)$$

- Local similarity measure: Here, the similarity of each feature of the reference and test speech feature vector is investigated. This formulation is shown in equation (12). The local distance is measured using the Euclidean distance presented in equation 12, where i, j in x_i and y_j denote reference feature x_i and test feature y_j. Here, we denote the similarity measure by the cost function as Euclidean distances. This means, the distance or similarity between the mfcc features of the ith test features and the jth reference features are computed.

$$d(x,y) = \sqrt{\sum_{i=1}^{k}(x_i - y_i)^2} \quad (12)$$

The similarity is measured the distance horizontally, vertically and diagonally. The horizontal, vertical and diagonal distances are denoted by $(i-1,j), (i,j-1), (i-1,j-1)$. The transition costs are also estimated either diagonally, or horizontally, or vertically. The DTW transition is shown in figure 4. The ith and j^{th} entries denote a minimum in a weighted Euclidean distance matrix.

An experimental result of the similarity measures using DTW is shown in figure 5. This shows the similarity measurement of the mfcc features of the reference speech and the test speech.

3. Second Scenario

Here, an automatic speech recognition (ASR) system is intended to be used for an assisted living. In a nursing home, elderly or disabled persons are unable to perform basic tasks. The ASR system assists the physically or disabled persons to perform certain tasks automatically. In this scenario, the nursing room is embedded with the ASR system. The user or the physically disabled person utters a command in the room which has to be performed. The speech recognition system preforms the command as asked. If the command is not possible to follow, there is a feedback to the user.

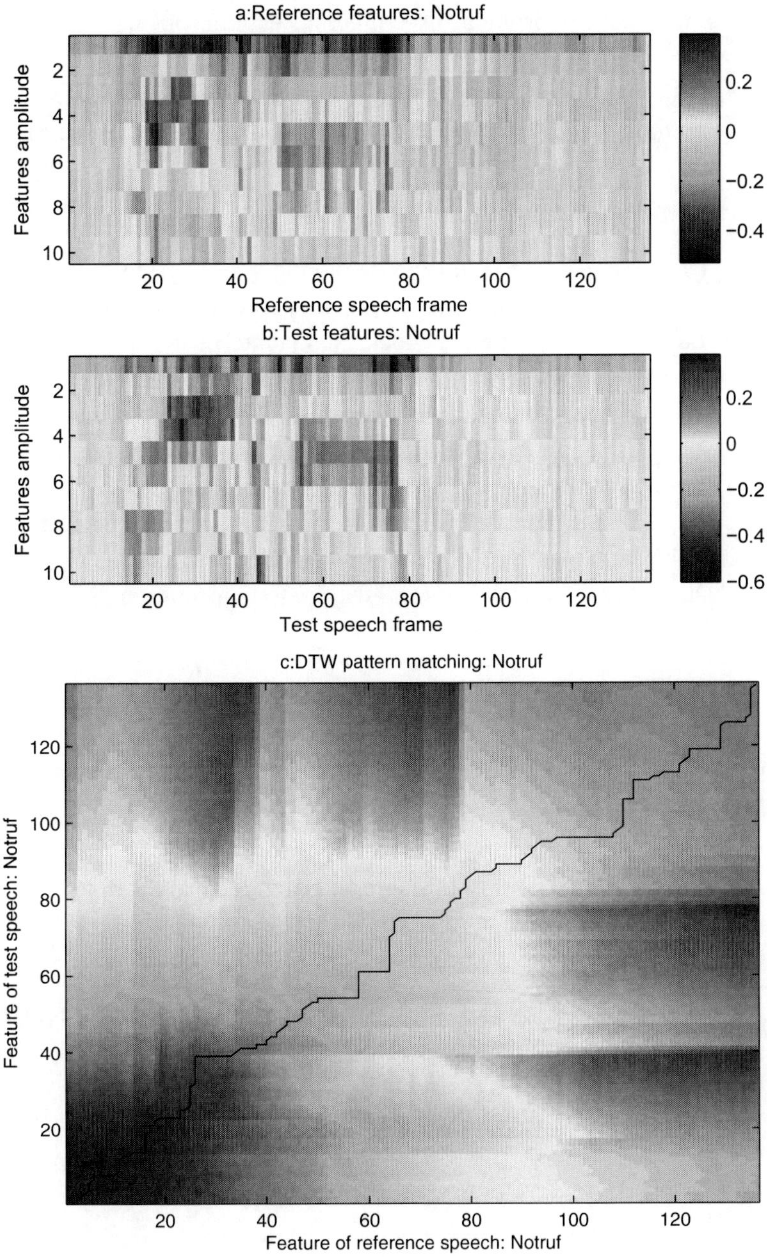

Figure 5. Pattern matching of reference and test speech: Notruf.

We approach this problem by collecting data according to our predefined list. We extract features applying shift invariant local trigonometric transformation (SILTT) and classify them in order to be recognized. The recognition is done by applying the DTW pattern matching tool discussed in section 2.2.

3.1. Feature Extraction: SILTT

The SILTT is an adaptive feature extraction technique. This contains multiple representations of the signal divided successively into smaller blocks and expanded into local bases in each block. The state of the art is also similar to the MFCC. Particularly in SILTT, windowing the signal applying window function and folding operation result in disjoint intervals. These are then used for spectral analysis by applying a discrete cosine transform typed IV i.e. DCT-IV. Finally, the features are the cost function which are the spectral entropy of the spectrums of the intervals. Then the inverse discrete cosine transform is computed following an unfolding operation. We have divided the section into spectral shaping, spectral analysis and feature transformation. Below, we briefly discuss the techniques.

3.1.1. Spectral Shaping

Here, we explain the windowing of the signal and the folding operator in order to obtain disjoint signal intervals.

Windowing the Signals: In this step, the signal is dyadically windowed into $I_j = [a_j, a_{j+1})$ by a window function $b_j(t)$ shown in equation (13). $b_j(t)$ uses a rising cut-off function $\beta(t)$. r denotes an overlapping.

$$b_j(t) = \begin{cases} \beta(\frac{t-a_j}{r}) & t \in [a_j - r, a_j + r) \\ 1 & t \in [a_j + r, a_{j+1} - r) \\ \beta(\frac{a_{j+1}-t}{r}) & t \in [a_{j+1} - r, a_{j+1} + r) \\ 0 & t \in (-\infty, a_j - r] \cup [a_{j+r}, \infty) \end{cases} \tag{13}$$

$$\beta(t) = \begin{cases} \sin\frac{\pi}{4}(1 + \sin\frac{\pi}{2}t) & \text{for } -1 \leq t \leq 1 \\ 0 & \text{for } t < -1 \\ 1 & \text{if } t > 1 \end{cases} \tag{14}$$

Folding Operator: This gives a smooth orthogonal projection of $b_j(t)$. The formulation of the folding operator is given in equation (15).

$$s_j[n] = \begin{cases} b_j[n]s[n] + b_j(2a_j - n)s(2a_j - n) & \text{if } a_j \le n \le a_j + r \\ s[n] \quad \text{if } a_j + r \le n \le a_{j+1} - r \\ b_j[n]s[n] + b_j[2a_{j+1} - n]s[2a_{j+1} - n] & \text{if } a_{j+1} - r \le n \le a_{j+1} \end{cases} \tag{15}$$

Now we have the disjoint signal interval which we use for spectral analysis applying DCT-IV discussed next.

3.1.2. Spectral Analysis: DCT-IV

Here we apply DCT-IV for spectral analysis and we compute the cost function of each finite length signal spectrums.

DCT-IV: The disjoint intervals obtained in the spectral analysis are multiplied by DCT-IV for their spectral analysis using equations (16) and (17).

$$\phi_k^j[n] = \frac{\sqrt{2}}{\sqrt{|I_j|}} \cos \frac{\pi}{|I_j|}[k + \frac{1}{2}][n - a_j]\chi_{I_j}[n] \tag{16}$$

$\chi_{I_j}[n]$: Characteristic function.

In this basis, we represent now the signal by equation (17).:

$$S_j[n] = \sum_{k \in \mathcal{Z}} c_k^j \phi_k^j[n] \quad \text{with} \quad c_k^j = \left\langle S_j[n], \phi_k^j(n) \right\rangle_{\mathcal{L}^2(\mathcal{R})} \tag{17}$$

In equation (17), $\left\langle S_j[n], \phi_k^j(n) \right\rangle$ is an inner product of $\mathcal{L}^2(\mathcal{R})$ i.e. the inner product of $S_j[n]$ and $\phi_k^j(n)$.

Cost Function: The spectrally transformed speech is then used to extract the features by computing the spectral entropy using equation (18) and equation (19). In equation (18), S is the spectrum computed by DCT-IV and S_k is the spectrum at k^{th} frequency components. In equation (18), S_i is the energy of the kth frequency component of the spectrum and $\breve{s}_k$ is the probability mass function (PMF) of the spectrum. The entropy is computed by equation (19).

$$\breve{s}_k = \frac{S_k}{\sum_{k=1}^{N} S_k} \quad \text{for} \quad k = 1, 2, \cdots, N \tag{18}$$

$$H(\breve{s}) = -\sum_{\breve{s}\in S} \breve{s}_k \log_2(\breve{s}_k) \tag{19}$$

3.1.3. Feature Transformation

Here, the signal is used to apply the inverse DCT-IV and the unfolding operation in order to generate speech features. The inverse DCT-IV is the same as DCT-IV given in equation (16).

Unfolding Operation: In the feature transformation, the spectrum of the finite length signal is $y_j[n]$ shown in equation (20) obtained by applying the unfolding operation. This undos the operation of the folding operation.

$$y_j[n] = \begin{cases} b_j[n]y[n] - b_j[2a_j - n]y_j[2a_j - n] & \text{if } a_j \leq n \leq a_j + r \\ y_j[n] \quad \text{if} \quad a_j + r \leq n \leq a_{j+1} - r \\ b_j[n]y_j[n] + b_j[2a_{j+1} - n]y[2a_{j+1} - n] & \text{if } a_{j+1} - r \leq n \leq a_{j+1} \end{cases} \tag{20}$$

In figure ??, we see the reference and test features extracted by SILTT feature extraction technique in a and in b, where, in c we see the SILTT features matching by DTW pattern matching tool.

4. Third Scenario

In a workshop, a lab or an industrial environment, the noise is quite common and it has a rich variety. This can be mild, strong, steady and varying. The noise problem can be solved or at least diminished efficiently and economically by using a speech recognition system. Here, the speech recognition is a replacement of the human listener by a machine. The commands between a human being and a computer can be performed remotely in order to operate the machine. Hence a human being is not required to expose the noisy environment frequently and thus healthy potentials are preserved. At the same time, it is economically beneficiary because the man power is replaced by a computer.

The solution is done by solving the noise problem first and then recognizing the enhanced speech. Subsequently one embeds the system into the machine in order to function itself as a replacement for a human being. A rough overview of three types can be seen in figure 6. There we see:

- Mild noise is modeled by white Gaussian noise.

- Time varying noise is modeled by a Gaussian process.
- Strong noise is modeled a Poisson process

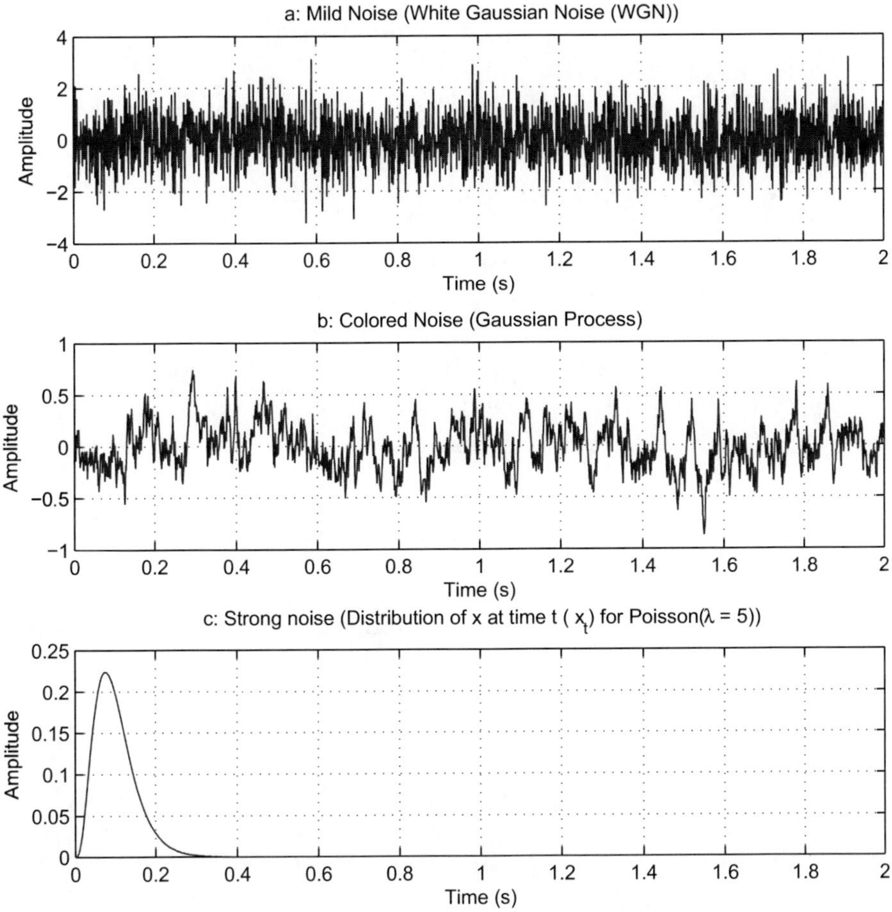

Figure 6. Hybrid noise formulations.

4.1. Strong Noise Solution: Matched Filter

Because of large amplitude, short time duration, randomly and rarely occurring events, the strong noise is modeled by shot. Their random characteristics are described by the Homogeneous Poisson distribution. The formulation of strong noise is given in equation (21). There, $d[n]$ is an impulse event having amplitude a_i and this appears at Poisson points n_i. Now for a short timed interval the shots may occur or may not occur. Thus, the probablity of its occurrence is 1 or 0.

$$d[n] = \sum_i a_i \delta[n - n_i] \tag{21}$$

Now we have the observation o mixed with the signal s and noise d. This is formulated in equation (22). Here, $s[n]$ is assumed to be an autoregressive process (AR).

$$o[n] = s[n] + d[n] \tag{22}$$

$s[n]$ is modeled by a linear prediction analysis as shown in equation (22). The solution to the coefficients is obtainted by Yule-Walker approach. Since equation (23) is a predicted estimate of $s[n]$, therefore $s[n]$ and the prediction $\hat{s}[n]$ may not be the same and there is an error which is the difference between $s[n]$ and $\hat{s}[n]$. This error is white noise $z_b[n]$.

$$\hat{s}[n] = \sum_{i=1}^{p} \alpha_i s[n - i] \tag{23}$$

Thus, whitening the signal $o[n]$ gives us white noise $z_b[n]$ and shots $z_d[n]$ as $a_i h[n - n_i]$. This is shown in figure 7. In this figure, $F(z)$ is a whitening filter and $F(z) = 1 - \sum_{i=1}^{p} \alpha_i z^{-i}$. Now, the shots are treated by a matched filter based on the theoretical approach of statistical detection. In this case, it supports the following two hypotheses:

- Hypothesis H_0: Noise is present only for $a_i = 0$.
- Hypothesis H_1: $\delta[n]$ with noise for $a_i \neq 0$.

We also have $\theta = \sigma^2 \sqrt{(1 + \alpha_1^2 + \alpha_2 + \cdots + \alpha_p^2)}$ and σ^2=variance of white noise.

According to statistical detection theory, the matched filter is the best detector when it is a time reversed version of $F(z)$ and this time the signal to noise ratio (SNR) is maximum. Thus the matched filter is given in equation (24).

$$\text{Matched filter:} \quad G(z) = F(-z) \quad \text{at} \quad H_0 \quad \text{and SNR is maximal.} \tag{24}$$

The results of application of the matched filter for the shots detection can be seen in figure 8 where the matched filter detected shots can be seen at close to samples at 100 and close to samples at 500.

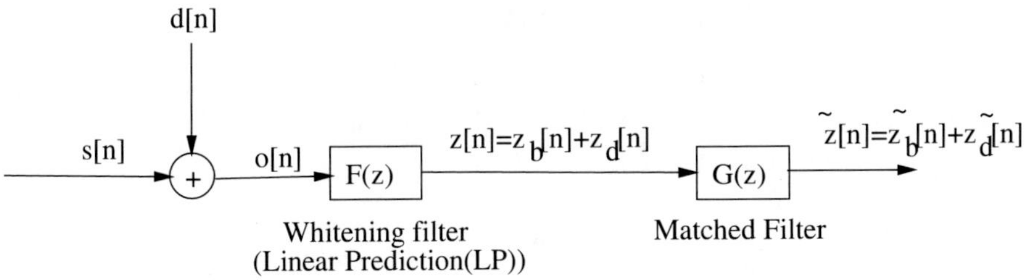

Figure 7. Matched filter for Shot Noise.

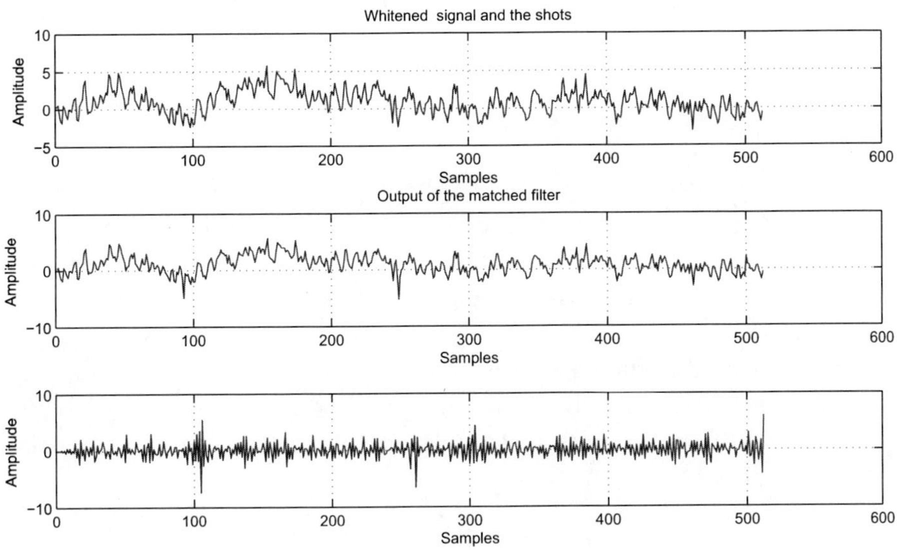

Figure 8. Strong noisy signal and output of matched filter.

4.2. Time Varying Noise Solution

For time varying noise solution, we have the observation y mixed with signal s and noise b shown in equation (25). Both signal and noise are modeled by linear prediction (LP) analyses.

$$y[n] = s[n] + b[n] \tag{25}$$

The signal s is modeled by equation (26) and the noise b is modeled by equation (27). In the equations, α_i and β_i are unknown signal and noise model coefficients which have to be determined, p and q are signal and noise model orders, g_s and g_b are weights, u_s and u_b are white noises.

$$s[n] = \sum_{i=1}^{p} \alpha_i s[n-i] + g_s u_s[n] \tag{26}$$

$$b[n] = \sum_{i=1}^{q} \beta_i b[n-i] + g_b u_b[n] \tag{27}$$

The solution to the signal model coefficients α is obtained by unconstrained least squares (ULS) and the solution to the noise model coefficients are obtained by the Yule-Walker equation.

4.2.1. Linear Prediction Analysis

In this approach, the signal is considered as an auto-regressive (AR) process. The parameters of this process are approximated by applying linear prediction (LP) analysis by using equation (28). In LP analysis, the current speech sample $s[n]$ is predicted is using linear combinations of the past p many samples with the coefficients α_i.

The predictor coefficients of LP are α_i for $i = 1, 2, \cdots, p$. This is shown in equation (28); there the predicted signal of the actual s is denoted by $\hat{s}$. In equaion (28), model order $\approx 2 \times$ (Number of formants) $+ 4$. A formant is an acoustical event.

$$\hat{s}[n] = \sum_{i=1}^{p} \alpha_i s[n-i] \tag{28}$$

The best approximation of the coefficient is the one which gives least error $e[n]$ schown in equation (29).

$$e[n] = s[n] - \hat{s}[n] = s[n] - \sum_{i=1}^{p} \alpha_i s[n-i] \tag{29}$$

However, in the least squares, we look for the minimum value of the mean squared error written in equation (30) where E is the expectation.

$$\eta = E\{e^2\} = \frac{1}{N}\sum_{n=0}^{N-1}[s[n] - \hat{s}[n]]^2 \tag{30}$$

Minimization of η'_k with respect to coefficients α_k results in equation (31). This gives p many linear equations shown in equation (32).

$$\eta'_k = \frac{\partial[E\{e^2\}]}{\partial \alpha_k} \tag{31}$$

$$r[i] = \sum_{k=1}^{p} \alpha_k r[i,k] \quad \text{for} \quad i = 1, 2, \cdots, p \tag{32}$$

The formulations of $r[i]$ and $r[i,k]$ are given in equations (33) and (34). Equation (33) is known as the signal energy at lag 0 which is also a correlation between signals $s[n]$ and $s[n-i]$ at lag 0.

$$r[i] = E\{s[n]s[n-i]\} \tag{33}$$

Equation (34) is a correlation between signals $s[n-i]$ and $s[n-k]$

$$r[i,k] = E\{s[n-i]s[n-k]\} \tag{34}$$

Now the solution to the coefficients of α is shown in equation (35). The corresponding matrix vector form of the coefficient vector α, the matrix $\mathbf{\Phi}$ and the correlation vector $\mathbf{r}$ are also shown below. The matrix $\mathbf{\Phi}$ is also termed as a correlation matrix. Equation (35) is known as the Yule-Walker equation. The Yule-Walker approach is also known as an auto-correlation approach. The inverse of $\mathbf{\Phi}^{-1}$ is solved by the Levinson-Durbin algorithm.

$$\alpha = \mathbf{\Phi}^{-1}\mathbf{r} \tag{35}$$

$$\mathbf{\Phi} = \begin{bmatrix} r[0] & r[1] & \cdots & r[p-2] & r[p-1] \\ \vdots & \vdots & \vdots & \vdots & \vdots \\ r[p-1] & r[p-2] & \cdots & r[1] & r[0] \end{bmatrix}; \alpha = \begin{bmatrix} \alpha_1 \\ \vdots \\ \alpha_p \end{bmatrix}; \mathbf{r} = \begin{bmatrix} r[1] \\ \vdots \\ r[p] \end{bmatrix}$$

Equation (35) is used to obtain the coefficients of our noise model.

4.2.2. Signal Model: Modified Covariance Approach (ULS)

The signal is modeled by the modified covariance approach. This is known as the unconstrained least squares (ULS) approach. This approach is briefly introduced here. In this approach, the forward linear prediction (FLP) analysis and the backward linear prediction (BLP) analysis are considered. Here, the minimization of forward linear prediction (FLP) error $e_p^f[n]$ and backward linear prediction (BLP) error $e_p^b[n]$ are considered. α_p^{fb} is the model coefficient computed in combination of FLP and BLP. The FLP error and the BLP error are written in equations (36) and (37). The computation of FLP and BLP does not include new statistical information but a consideration of more points from these two types of analyses generates an improved approximation of AR parameters.

$$\mathbf{e}_p^f[n] = \mathbf{s}_p^T[n]\alpha_p^{fb} \tag{36}$$

$$\mathbf{e}_p^b[n] = \mathbf{s}_p^T[n]\mathbf{J}\alpha_p^{fb} \tag{37}$$

$$\mathbf{J} = \begin{bmatrix} 0 & 0 & \dots & \cdots & 1 \\ 0 & 0 & 0 & \dots & 0 \\ \dots & \dots & \dots & \dots & \dots \\ 0 & 1 & 0 & \dots & 0 \\ 1 & 0 & 0 & \dots & 0 \end{bmatrix}; \mathbf{s}_p[n] = \begin{bmatrix} s[n] \\ s[n-1] \\ \vdots \\ s[n-p+1] \\ s[n-p] \end{bmatrix}; \alpha_p^{fb} = \begin{bmatrix} \alpha_p[1] \\ \alpha_p[2] \\ \vdots \\ \alpha_p[p-1] \\ \alpha_p[p] \end{bmatrix}$$

The combination of BLP and FLP errors is denoted by η_p^{fb}. This is computed by equation (38).

$$\eta_p^{fb} = \frac{1}{N} \sum_{n=p+1}^{N} [e_p^{f^2}[n] + e_p^{b^2}[n]] \tag{38}$$

The solution to the model coefficient α_p^{fb} is in equation (39) where $\mathbf{\Phi}_p^{-1}$ is obtained by fast covariance QR factorization. The matrix $\mathbf{\Phi}$ is not Toeplitz and therefore the Levinson-Durbin algorithm is not applied to compute the inverse of the matrix. In Yule-Walker approach and in Burg approach, the matrix $\mathbf{\Phi}$ is Toeplitz. Both approaches use the Levinson-Durbin algorithm for model coefficients. The solutions of the prediction coefficients are optimized each time to get ULS model coefficients where in Bark approach, the prediction coefficients are optimized only once. Therefore the signal model is called the unconstrained least squares. We have used the ULS approach to obtain the signal model coefficients [15].

$$\alpha_p^{fb} = \mathbf{\Phi}_p^{-1}\left[\eta_p^{fb}, \mathbf{0}_p\right]^T \quad \text{and } \mathbf{\Phi}_p = \sum_{n=p+1}^{N}\left(\mathbf{s}_p[n]\mathbf{s}_p^T[n] + \mathbf{J}\mathbf{s}_p[n]\mathbf{s}_p^T[n]\mathbf{J}\right) \quad (39)$$

Now we have the signal and the noise model coefficients. In this time varying noise solution, the signal is divided into sub-bands by the cosine modulated pseudo quadrature mirror filter band. In figure 9, the signal is subdivided into M-bands by the analysis filter 40 and the signals in each sub-bands are then used for spectral weighting in particular spectral minima tracking. Then Kalman filters are used in each sub-band before synthesizing the signals to x by a synthesis filter shown in equation (41). The whole approach can be seen in figure 9. In this figure, the sampling rate conversion is not shown in order to keep the figure simple. Here, we describe the M-bands Kalman filter in brief. The details of this can be found in [2], [3], [4], [5].

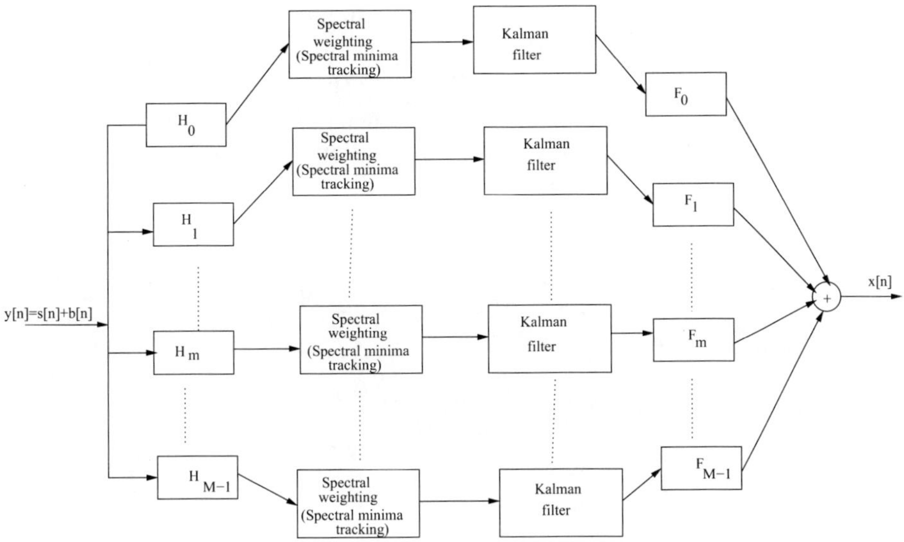

Figure 9. M-band Kalman filter for colored noise problem.

4.2.3. Sub-bands Decomposition

The type of sub-band is called cosine modulated pseudo QMF M-band banks (Pseudo QMF). The impulse response of the analysis filter $H_0(z)$ is $h_m[n]$ and it is shown in equation (40). There, $w[n]$ is L sampled Kaiser window, $M = 32$ and $\Omega = (-1)^m \frac{\pi}{4}$, $m = 0, 1, \cdots, M-1$ and the sampling rate conversion is critical.

$$h_m[n] = 2w[n]\cos\{\frac{\pi}{M}(m+0.5)(n-\frac{L-1}{2}) + \Omega_m\} \tag{40}$$

Similarly, the impulse response of synthesis filter $F_m(z)$ is $f_m[n]$ as shown is equation (41).

$$f_m[n] = 2w[n]\cos\{\frac{\pi}{M}[(m+0.5)(n-\frac{L-1}{2}) - \Omega_m\} \tag{41}$$

4.2.4. Spectral Minima Tracking in Sub-bands

Here, the goal is to suppress noise by tracking the local minimum of the noisy speech using equations (42) and (44). In the equations, $P_{s_m}(k)$ is the estimate of the noisy speech power spectral density (psd) at band m. $|Y_m(k)|$ is the noisy signal spectrum at band m and frequency k, $P_{b_m}(k)$ is the noise spectral density of k at band m, and $\rho = 0.7$, $\tau = 0.96$, and $\upsilon = 0.998$: These are selected experimentally [1].

The following equations describe the relations between the used concepts.

$$P_{s_m}(k) = \rho P_{s_{m-1}}(k) + (1-\rho)|Y_m(k)|^2 \tag{42}$$

If

$$P_{b_{m-1}}(k) < P_{s_m}(k) \tag{43}$$

$$P_{b_m}(k) = \upsilon P_{b_{m-1}}(k) + \frac{1-\upsilon}{1-\tau}(P_{s_m}(k) - \tau P_{s_{m-1}}(k)) \tag{44}$$

Else

$$P_{b_m}(k) = P_{s_m}(k) \tag{45}$$

Now, we briefly describe the Kalman filter applied in the sub-bands.

4.2.5. Kalman Filter for Colored Noise in Sub-bands

Here we provided the state-space formulations of the Kalman filter for colored noise in sub-bands. Now the state vector and the noise vector are s and $\mathbf{b}$. There, the state-space formulations are given below.

The first two equations describe the signal and the noise.

$$\mathbf{s}[n+1] = \mathbf{A}[n]\mathbf{s}[n] + \mathbf{g}_s\mathbf{u}[n] = \gamma^T[n]\mathbf{s}[n] \tag{46}$$

$$\mathbf{b}[n+1] = \mathbf{D}[n]\mathbf{b}[n] + \mathbf{g}_b\nu[n] = \psi^T[n]\mathbf{b}[n] \tag{47}$$

For further handling, it is useful to describe the equations in more detail.

$$\mathbf{s}^T[n] = \big[s[n-1], \cdots, s[n-p]\big];\ \mathbf{b}^T[n] = \big[b[n-1], \cdots, b[n-q]\big]$$

$$\mathbf{g}_s^T = \big[0, \cdots, 1\big]_{p\times 1}; \mathbf{g}_b^T = \big[0, \cdots, 1\big]_{q\times 1}; \mathbf{u}[n], \nu[n] : \text{White noise}$$

$$\gamma^T = \big[0, \cdots, 1\big]_{p\times 1}; \psi^T = \big[0, \cdots, 1\big]_{q\times 1}$$

$$\mathbf{A}[n] = \begin{bmatrix} \mathbf{0}_{p-1\times 1} & \mathbf{I}_{p-1\times p-1} \\ \alpha_{\mathbf{p}}^{\mathbf{T}} & \end{bmatrix}_{p\times p}; \mathbf{D}[n] = \begin{bmatrix} \mathbf{0}_{q-1\times 1} & \mathbf{I}_{q-1\times q-1} \\ \beta_{\mathbf{q}}^{\mathbf{T}} & \end{bmatrix}_{q\times q}$$

Now, the observation vector is $\mathbf{y}$.

$$\mathbf{y}[n+1] = \mathbf{C}[n]\mathbf{y}[n] + \mathbf{G}\mathbf{w}[n] \tag{48}$$

$$\mathbf{C}[n] = \begin{bmatrix} \mathbf{A}[n] & \mathbf{0} \\ \mathbf{0} & \mathbf{D}[n] \end{bmatrix}, \mathbf{y}[n] = \begin{bmatrix} \mathbf{s}[n] \\ \mathbf{b}[n] \end{bmatrix}, \mathbf{G} = \begin{bmatrix} \mathbf{g}_s & \mathbf{0} \\ \mathbf{0} & \mathbf{g}_b \end{bmatrix}, \mathbf{w}[n] = \begin{bmatrix} \mathbf{u}[n] \\ \mathbf{v}[n] \end{bmatrix}$$

$$\text{Observation:} y[n] = \mathbf{H}[n]^T\mathbf{y}[n] = \big[\gamma[n] \quad \psi[n]\big] \begin{bmatrix} \mathbf{s}[n] \\ \mathbf{b}[n] \end{bmatrix} \tag{49}$$

The prediction equations are now:

- $\hat{y}[n|n-1] = \mathbf{C}[n]\hat{y}[n-1|n-1]$
- $\mathbf{P}[n|n-1] = \mathbf{C}[n-1]\mathbf{P}[n-1|n-1]\mathbf{C}^T[n-1] + GW[n]G^T$

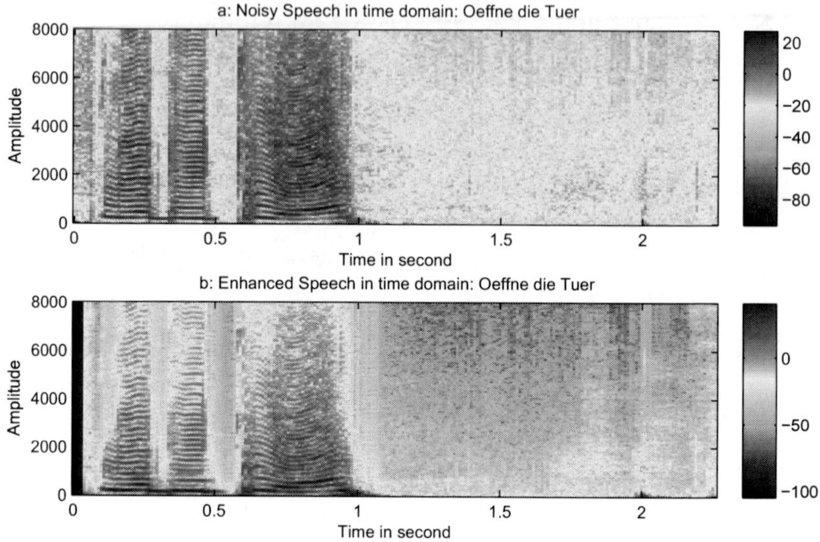

Figure 10. Speech is enhanced by M-band Kalman filter: Oeffne die Tuer

Next we bring this in relation with the Kalman gain what will provide us the updating. The Kalman gain is:

$\mathbf{K} = \mathbf{P}[n|n-1]\mathbf{H}[n-1](\mathbf{H}^T[n-1]\mathbf{P}[n|n-1]\mathbf{H}[n-1])^{-1}$ $\mathbf{P}[n|n] = (\mathbf{I}-\mathbf{K}[n]\mathbf{H}^T[n])\mathbf{P}[n|n-1]$

Updating is now shown by the next two equations. $\hat{\mathbf{y}}[n|n] = \hat{\mathbf{y}}[n|n-1] + \mathbf{K}[n](\mathbf{y}[n] - \mathbf{H}^T[n-1]\hat{\mathbf{y}}[n|n-1])$ $\mathbf{P}[n|n] = (\mathbf{I}-\mathbf{K}[n]\mathbf{H}^T[n])\mathbf{P}[n|n-1]$ In figure 10, we see the result of M-bands Kalman filter. In a, we see the noisy speech signal in the time domain and the enhanced speech signal in b. The signal to noise ratio of the signal (SNR) is 26.78.

4.3. Feature Clustering, Learning and Classification

Here, we have used a clustering approach in order to classify the feature vectors. The clustering methods in machine learning are unsupervised. They depend on a similarity measure. Below we describe them in brief.

4.3.1. Clustering Approach

The clustering methods in machine learning are unsupervised. They depend on a similarity measure. The clustering splits the feature sets into different sets. The conditions for the splitting are obtained by the similarity measure. They say that the objects in one set are close together and objects in different sets are distant. Features are represented in terms of the parameters of the statistical model for the states. The learning is an estimation of the best parameters for the best estimation, also to find the best centers of the clusters. Testing is to control the results using the testing set. The GMM parameters are initialized K-means algorithm. Below, we briefly explained the K-means clustering approach and the GMM classification.

K-means Clustering: The main characteristic of K-means is that the number of clusters denoted by K is fixed in advance. Each cluster set has a center and it should contain just the objects that are closest to the center. The classes are the predefined list of spoken commands. In this clustering, K is the number of classes for $k = 1, \cdots, K$. If u_k are feature vectors, we have to define arbitrarily as k possible centers and define the clusters by the distance to u_k. We need to compute centers of the clusters and replace u_k. We compute new clusters until no change is observed in shuffling the features to clusters. These are our clusters and these are used to initialize GMM parameters.

4.3.2. Feature Classification and Model: GMM

The features are modeled by G many Gaussian mixtures and $g = 1, 2, \cdots, G$. $\mathbf{c}_g$ is a mixture weight, μ is a mean vector and Σ is a covariance matrix.

$$p(\mathbf{o}_t|q) = \sum_{g=1}^{G} c_g p(\mathbf{o}_t \mid \mu_g, \mathbf{\Sigma}_g) = \sum_{g=1}^{G} c_g \mathcal{N}(\mathbf{o}_t; \mu_g, \Sigma_g) \tag{50}$$

$$\mathcal{N}(\mathbf{o}_t; \mu_g, \Sigma_g) = \sum_{g=1}^{G} \{\mathbf{c_g} \frac{1}{(2\pi)^{l/2} \mid \Sigma_g \mid^{1/2}} e^{\left(-\frac{1}{2}(\mathbf{o}_t - \mu_g)^T \Sigma_g^{-1} (\mathbf{o}_t - \mu_g)\right)}\}$$

$$\mathbf{b}_j(\mathbf{o}_t) = \sum_{g=1}^{G} c_g \frac{1}{(2\pi)^{l/2} \mid \Sigma_g \mid^{1/2}} \exp\big(-\frac{1}{2}(\mathbf{o}_t - \mu_g)^T \Sigma_g^{-1} (\mathbf{o}_t - \mu_g)\big) \tag{51}$$

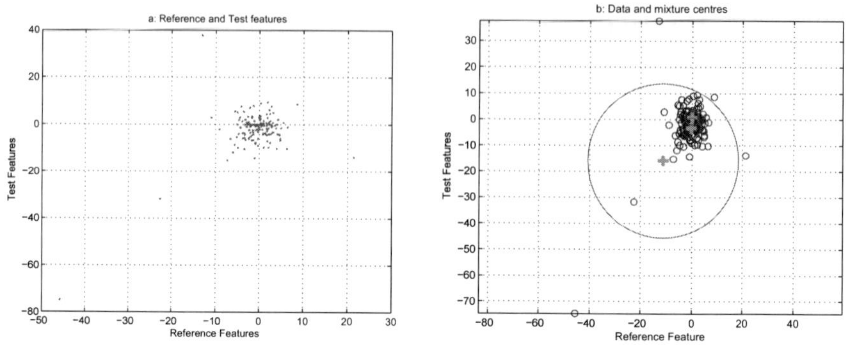

Figure 11. SILTT reference and test features and related GMM parameters: Öffne die Tür

GMM parameters are optimized by Expectation-Maximization (EM) algorithm. This is used for the recognition of our spoken commands in the hybrid noisy environment where the noises are occurring simultaneously.

In figure 11, in a, we see the reference SILTT features and the test SILTT features. In a, we see the features drawn from the same class, namely from the German spoken command: "öffne die Tür". Therefore, we see only one cluster. The clusters, the classes are some common terms in clustering, classification, learning and in testing. For this, we need to use more data, and the goal of the clustering is to represent the classes. Since we have used only one class, we see only one cluster in a in figure 11. In the same figure in b, we see related GMM parameters namely mixtures weights, mean vector and covariance matrix. These mixture weighs are prior but they sum up to 1. These are selected randomly. In this example, the GMM parameters such as the mean vector and the covariance matrix are obtained from the test and reference speech features using the features of two speech signals and these are shown in b in figure 11. The clustering and the classification are types of recognition tasks. We have not commented on the recognition because we used here only two signals. In order to comment on the clustering or classification recognition for the speech recognition problem, more training should be considered. This extensive evaluation is one of our future task. For this implementation, we have used the Netlab toolbox [19]. In figure 12, we see the probability density function (pdf) of the reference SILTT features and the test SILTT features using GMM. We see only one mode in the GMM model because we have only two signals from the same

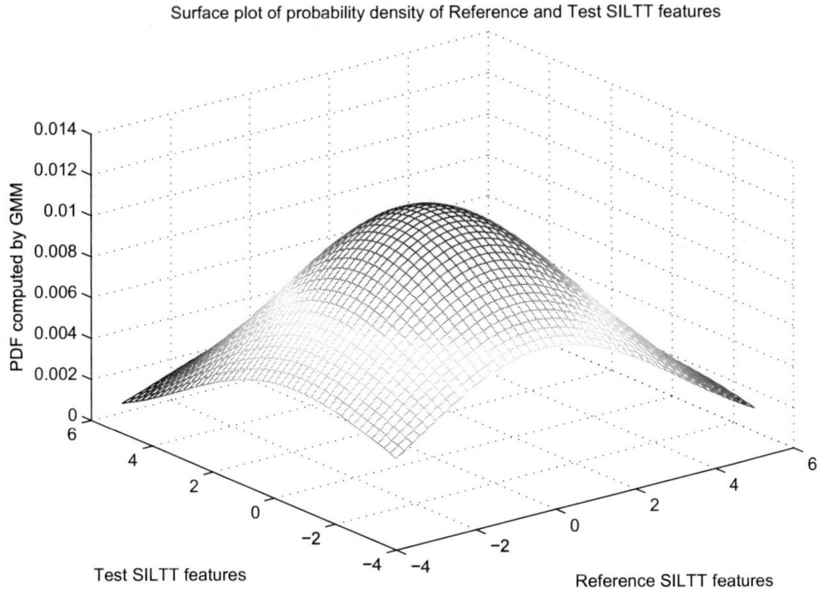

Figure 12. Probability density function (pdf) of reference and test SILTT features using GMM : Oeffne die Tuer

spoken command i.e. the same class. In the figure, we first computed the SILTT features of the signals and then we provided the pdf of the GMM of the feature sets of the reference and test speech. We term one speech signal as reference and the other is test. Normally these terms are used when more training and test sets are used. Our example can easily be extended for more training and test sets. The recognition accuracy of this has to be computed using these training and testing data sets. This is not shown here. The example shown in figure 12 is implemented using Netlab toolbox [19].

5. Conclusion

In this chapter we have provided the communication between a human and a machine in a hybrid noisy environment. We have considered three different types of noise. We have provided three different scenarios such as medical ward, health support center and industrial hybrid noisy environment where the three types of noise can take place. We have provided mathematical details of the model including our experimental results using a small amount of data. For reliability more training is suggested. In future we will extend this in detection based hybrid noisy speech recognition system where we plan to use more sophisticated psychoacoustics analysis for perceptual speech feature extraction and recognition of these features using more successful statistical pattern recognition tool Hidden Markov model (HMM) based on GMM classification.

References

[1] Gerhard Doblinger. *Computationally Efficient Speech Enhancement by Spectral Minima Tracking in Subbands*, Eurospeech, 1995

[2] Frederic Mustiere, Miodrag Bolic, Martin Bouchard. *Improved Colored Noise Handling in Kalman-Based Speech Enhancement Algorithms, IEEE*, ICASSP, 2008.

[3] Saeed V. Vaseghi. *Multimedia signal processing: Theory and Applications in Speech, Music, and Communications*, Wiley, USA, 2008.

[4] Yuan-Pei Lin, P. P. Vaidyanathan. *A Kaiser Window Approach for the Design of Prototype Filters of Cosine Modulated Filterbanks, IEEE*, March 1998, vol 5, Num 6.

[5] Charles K. Chui, Guanrong Chen. *Kalman Filtering with Real-Time Applications*, Springer, Berlin, 1987.

[6] Xuedong Huang, Alejandro Acero, Alex Acero, Hsiao-Wuen Hon. *Spoken Language Processing: A Guide to Theory, Algorithm and System Development*, Prentice Hall,1999.

[7] Sungwook Chang, Y. Kwon, Sung-il Yang. Speech Recognition System using Adapted Local Trigonometric Transformation, *IEEE-SP International Symposium*, Pittsburgh, PA, 1998.

[8] Gerard Blanchet, Maurice Charbit. *Digital Signal and Image Processing using Matlab*, ISTE, 2006.

[9] David Bies. *Engineering Noise Control : Theory and Practice*,Taylor & Francis, 2009.

[10] Sheuli Paul, Michael Richter, Steven Liu. *Hybrid Solution to Single- channel Hybrid Noisy Speech for an Industrial Environment, ISSPIT,IEEE*, 2012.

[11] Sheuli Paul, Michael Richter. *Human Speech Recognition (SR) and a new Feature Extraction Technique for SR System*, Vol 9, MLDM, 2013.

[12] Sheuli Paul, Michael Richter. *A Dynamic Automatic Noisy Speech Recognition (DANSR) system for a Single-channel Hybrid Noisy Industrial Environment* , Vol. 21. ICA, June 2013.

[13] P.C. Yip. *The Transformation and Applications Handbook*, CRC Press, 2nd Ed, February, 2000.

[14] S. B. Davis, P. Mermelstein. *Comparison of Parametric Representations for Monosyllabic Word Recognition in Continuously Spoken Sentences, IEEE ASSP*, No. 4, Vol. 28, pp. 357-366, 1980.

[15] Larry Marple. *Digital Spectral Analysis with Application, IEEE ASSP*, 1980.

[16] Eva Wesfried, Mladen Victor Wickerhauser. Adapted Local Trigonometric Transforms and Speech Processing, *IEEE Transactions on Signal Processing*, No. 12, Vol. 41,1993

[17] Mladen Victor Wickerhauser. *Adapted Wavelet Analysis from Theory to Software, IEEE Press*, 1994.

[18] Lawrence R. Rabiner, Biing-Hwang Juang. *Fundamentals of Speech Recognition*, Prentice Hall, 1993.

[19] Ian Nabney. *NETLAB: Algorithms for Pattern Recognition*, Springer, 2002.

In: Noise Pollution
Editor: Eugene Roberto Nicchi
ISBN: 978-1-63321-110-0

Chapter 4

EVALUATION OF EFFECTS ON WORKERS' HEALTH DUE TO EXPOSURE TO DIFFERENT INTENSITY SPLS ARISING FROM VARIOUS WORK ACTIVITIES: STANDARDIZED PROCEDURES AND INTEGRATIVE METHODS

***Adolfo Sabato**, *Alessandro Sabato*† and *Alfredo Reda*‡**
Laboratorio di Tecnica del Controllo Ambientale, DIMEG,
Università della Calabria, Italy

ABSTRACT

Risks arising from physical agents - such as noise pollution - are sometimes underrated or not treated in a suitable way. This happens especially when workplace wellbeing is considered. In Europe, security is ensured with the *Directive 2003/10/EC of the European Parliament and of the Council of February 6th, 2003 on the minimum health and safety requirements regarding the exposure of workers to the risks arising from physical agents (noise)*. It supplies the minimum health requirements to assure health-care and protection from loud sounds to those workers who are exposed to noise as a result of their work. Nevertheless, this standard

* email: labtca@unical.it
† email: sabatoale@gmail.com
‡ email: alfredoreda@gmail.com

is far away from guaranteeing a complete safeguard because of several shortcomings.

In this chapter, the working activities of workers who remain in noisy workplaces for prolonged periods are analyzed. These activities may produce serious risks for worker's health with particular reference to the organ of hearing and its psychophysics functions. Therefore, parameters and acoustics quantities - which allow a better understanding of the physical phenomenon appointed within the Directive - are analyzed using data recorded with dosimeters and real time analyzers. Works carried out in environments characterized by different acoustic intensities are considered in this analysis. Depending on exposure time, each noisy environment can produces annoyance phenomena or permanent psychosomatic damage. In particular, the following activities are subject to investigation: office activities, work activities of public transportation drivers, university and research center's laboratories (e.g. mechanics labs, engine testing room, material stress testing labs, etc.) and industrial activities.

Based on the sampled data, acoustics parameters used within the current Directive are evaluated to highlight that it underestimates the effective noise to which workers are subjected. The same data is used to prove that these parameters cannot guarantee a protection from noise as effective as in the Directive's original aims. In practice, through several in-situ measurements the intensity of acoustic parameters is evaluated according to what is stated within the Directive. Then, the same data are analyzed using an integrative procedure – proposed by the authors – to evaluate the influence of impulsive components, which cannot be determined with the current standards. In addition, the influence of sound phenomena's spectral composition and its influence on the interested portion of human cochlea are achieved.

1. INTRODUCTION

Noise has become a prominent feature of the environment. Nowadays, people are constantly subjected to an increasing number of acoustic sources. Among them noise arising from: transportation, industrial activities, and work places. It has been demonstrated that noise interferes in complex task performance, modifies social behavior, and may cause several health problems (e. g.: annoyance, disturbance in sleep, interference with communication and other harmful effects) [1]. Noise can be defined as an invisible pollutant, completely different from other typologies of contaminant, and therefore unique in its characteristics. Indeed, the idea that noise pollution may have physical and health implications on people's life is quite recent [2].

The risks arising from physical agents - such as noise pollution - are sometimes underrated or not always treated in a suitable way, particularly when workplace well being is considered. Indeed, it is absolutely unreasonable that workers may get injured while performing their jobs. For this reason, many regulations have been proposed to protect workers from risks arising from noise. In Europe, security is ensured with the *Directive 2003/10/EC of the European Parliament and of the Council of February 6th, 2003 on the minimum health and safety requirements regarding the exposure of workers to the risks arising from physical agents (noise)* [3]. It supplies the minimum health requirements to assure health-care and protection from loud sounds to those workers who are exposed to risks from noise as a result of their work. While the Directive addresses problems coming from noise exposition and suggests practical requirements for protecting workers, it is far away from guaranteeing a complete safeguard because of certain shortcomings [4].

Following the introductory, in which an overview of the hearing organ and the main acoustic descriptors is given, a critical analysis of noise produced by different work is carried out. Activities made in environments characterized by different acoustic intensities are considered. Depending on exposure time, each environment can produce effects on the human health that range from annoyance phenomena to permanent psychosomatic damage [5]. In particular, the following activities are hereby subject to investigation: office activities, work activities of public transportation drivers, university and research center's laboratories (e.g. mechanics labs, engine testing room, material stress testing labs, etc.) and industrial sector's activities. Through several in-situ measurements the intensity of acoustic descriptors is evaluated according to what is stated within the Directive. Then, to prove that chosen parameters cannot guarantee a full-protection from risk arising from noise, the same data will be analyzed using an integrative procedure to evaluate both the influence of impulsive components and sound spectral composition, which cannot be properly determined when currently used procedures are applied.

2. Anatomy of the Human Hearing Organ

Exposition to noise can create harmful effects on human depending on the characteristics of the acoustic phenomenon (sound pressure levels, frequency, impulsive components, tonal components, etc.), exposition time, and subjective sensitivity of the person [6].

Hurtful consequences are mainly due to loud sound exposition, even for short time periods. From a clinical and pathological point of view, they consist of irreversible or not-completely reversible alterations of ear's anatomic functionalities [7]. On the other hand, exposition to lower and not excessively prolonged sound pressure levels (SPLs) can cause disturbance effects that lead to temporal alteration of the physical and mental conditions of the exposed person [8]. In addition, extremely sensitive subjects may address a feeling of discomfort even if they are exposed to very low SPLs [9]. This pathology is referred to as annoyance and it represents the subjective response to noise. Since it is a personal reaction - usually of physiological origin - it cannot be found through clinical analyses [1].

Before analyzing noise effects on workers' hearing organs, it is necessary to understand the structure of the ear and the way it transmits the received acoustic signals to the brain. As shown in Figure 1, the ear can be divided in three parts: outer, middle, and inner ear.

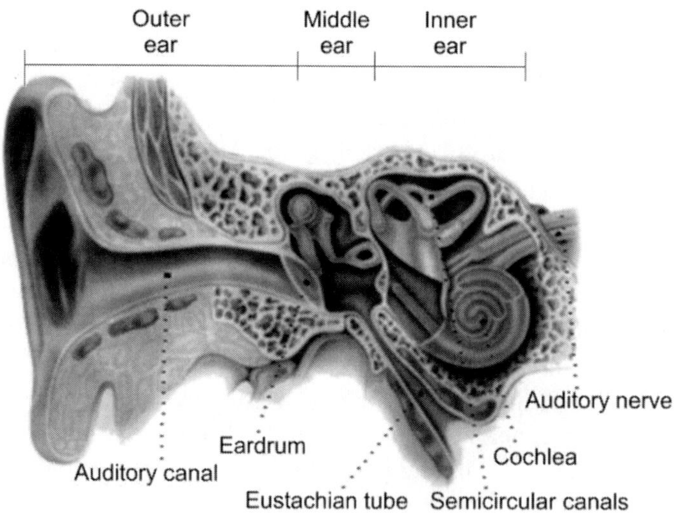

Figure 1. Section of the human hearing organ.

2.1. Outer Ear

The outer ear is the most external part of the organ. It includes the pinna (or auricle) and the external acoustic meatus (or auditory canal) [10]. The auricle collects the acoustic waves, which hit the surface of the ear, and conveys them to the inner part. The acoustic meatus stretches for nearly 24 millimeters and is surrounded with bones and cartilages. It ends on the tympanic membrane (or eardrum), which divides the outer ear from the middle one.

2.2. Middle Ear

Tympanum is a cone-shaped membrane fixed on the temporal bone through a fibro-cartilaginous ring called tympanic annulus. The middle ear, also referred to as tympanic cavity, is an air-filled, irregular-shape cavity, dug in the temporal bone. It includes three bones (or ossicles): the hammer (or malleus), incus (or anvil), and stapes (or stirrup) [10]. The first one is a small-elongated bone – measuring lengths from 7 to 9 millimeters. It is made of a handle, connected to the eardrum, a neck, and a head, connected to the incus. The hammer transmits vibration of the membrane produced by sound waves. The incus is connected with the stapes through its lower branch. The three bones transmit the sound from the tympanic membrane to the ventricles of the ear [11]. In Figure 2 a representation of the ossicles is shown.

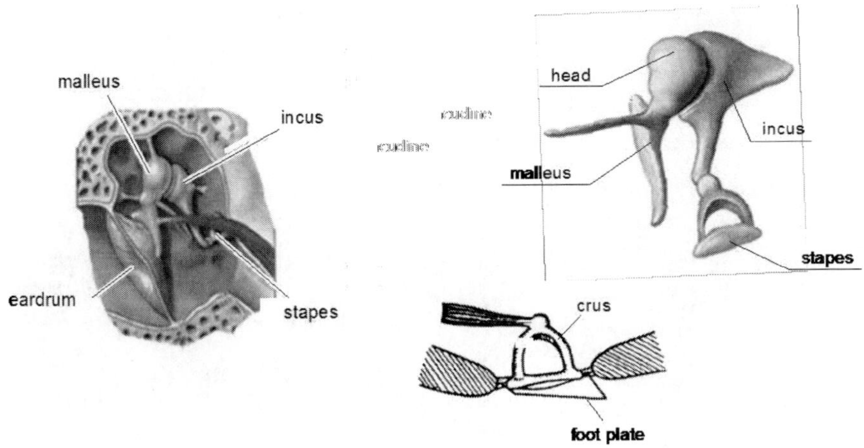

Figure 2. Representation of the three bones.

Within the bony canal, there are two muscles: the stapedius muscle and the tensor tympani [10]. The middle ear also contains a canal called Eustachian tube. It connects the tympanic cavity to the nasopharynx, the uppermost part of the pharynx. This canal compensates for the force acting on the external part of the tympanic membrane due to atmospheric pressure. The air, which enters through the Eustachian tube, balances the high stresses placed on the membrane resultant from the atmospheric pressure [11]. The ending part of the stapes, called the base, rests on the oval window and separates the middle ear from the inner ear.

2.3. Inner Ear

The inner ear contains the sensory organ for balance and hearing. Here, it is possible to distinguish the vestibule, three semicircular canals, and the cochlea. The vestibule is the construct of the ear that communicates with the middle ear through the oval window. The semicircular canals are made of two parts: a bony labyrinth and a membranous labyrinth contained within the bony labyrinth, which reproduces its morphology. The bone canals are connected to the vestibule through three orifices. These are located perpendicular to one another and are arranged in the upper and rear of the vestibule. The cochlea starts from the base of the vestibule, as a tubular space, then makes 2.5 turns around its axis [10]. Figure 3 shows the main components of the inner ear.

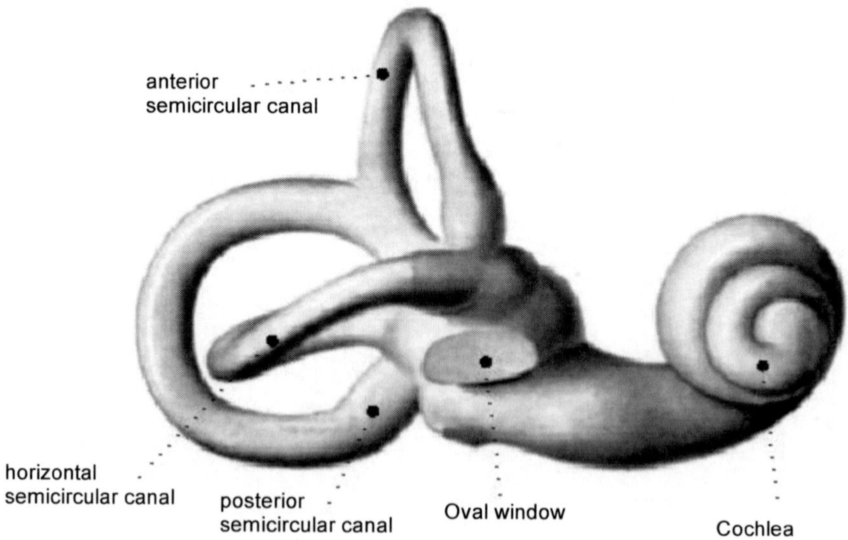

Figure 3. Components of the inner ear.

The cochlea is the auditory portion of the inner ear and is the core of the auditory system [12]. This coiled tube is divided into two portions (scale) by means of an inner membranous partition: the scala vestibule and the scala tympani (see Figure 4.a). The two scales are filled with a fluid called perilymph and are separated everywhere along their length but the top, where a connection is made through a hole called helicotrema. The helicotrema allows fluid to move within the two ramps [10]. The lengthwise partition that divides most of the cochlea is itself a fluid-filled tube, called scala media or cochlear duct. It is bounded on three sides by the basilar membrane, the stria vascularis, and Reissner's membrane (see Figure 4.b) and it is filled with a fluid called endolymph.

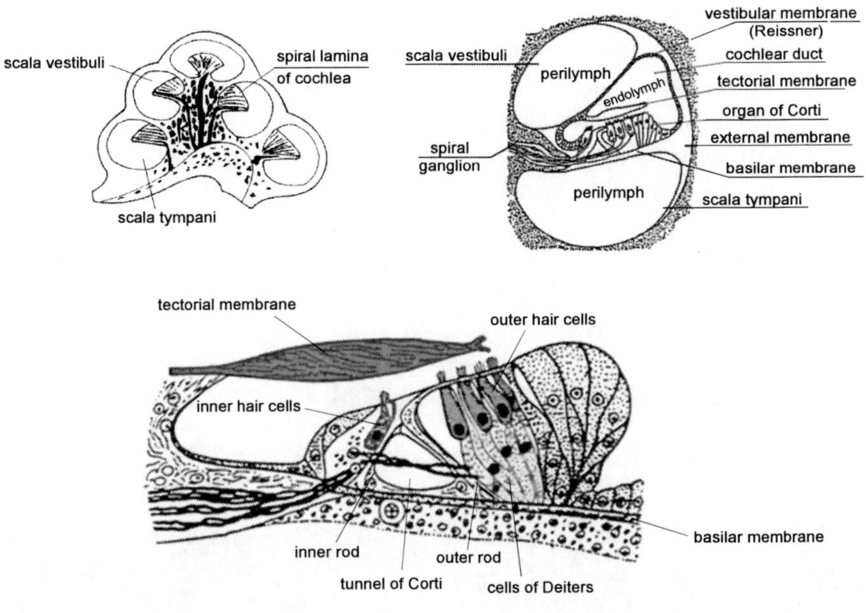

Figure 4. Sagittal section of the cochlea (a), of the cochlear canal (b), of the organ of Corti (c).

Sound vibration transmitted by the ossicular chain, becomes a change in the pressure of the fluid inside the cochlea. Because of this variation, the basilar membrane starts oscillating with different amplitudes depending on the intensity of the acoustic input. The basilar membrane is made of several elastic fibers, connected to two bony prominences. Fibers are short and dense at the beginning of the cochlea and become longer and thinned as they proceed toward the apex. At the same time, they modify their elasticity at an exponential rate. Different stiffness allows sounds of vary frequencies to excite different point of the membrane; areas close to the base are excited by high-frequency sounds, while areas close to the apex are excited by low-frequencies [13].

Resting on the basilar membrane and disposed along the entire cochlear duct, there is the Organ of Corti. As shown in Figure 4.c, it is made of several cells arranged in a pillar shape, which delimit the Tunnel of Corti. On both sides of that tunnel there are acoustic cells connected to cochlear nerve fibers (ganglion of Corti) [8]. There are two different types of cells: external and internal. The external cells are made of rigid hair and are connected to nerve endings called stereocilia. As shown in Figure 5, they are made of a bundle of tubes rigidly mounted at the top of the external cells and connected to the tectorial membrane, which transmits vibration from movements of the endolymph in the cochlear duct. The viscosity of the tectorial membrane has a fibro-gelatinous composition that allows it to adapt its shape to variations in the organ of Corti [14]. The internal cells, which, are connected to the basilar membrane have a sensing capacity which can transmit received signals to the cochlear nerve. Each module is made of a triplet of moveable hair cells and an internal cell. The physical structure of the system allows an amplification of high-frequency signals. Otherwise, this frequency range would be cut because of the reduced sensitivity of the motor cells.

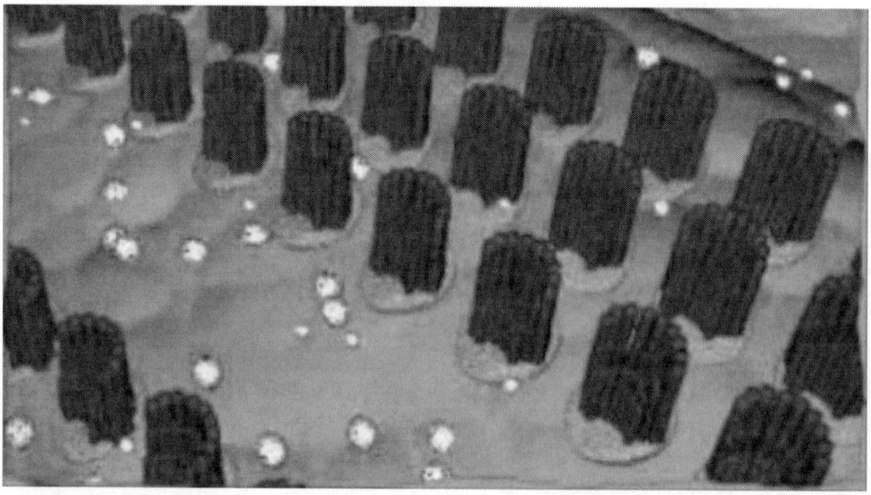

Figure 5. Stereocilia.

The mechanism underling the operation of the Organ of Corti is based on the electrical potential difference between the two scales because of the two different fluids running within them. The equivalent electrical circuit and electrical potentials are shown in Figure 6.

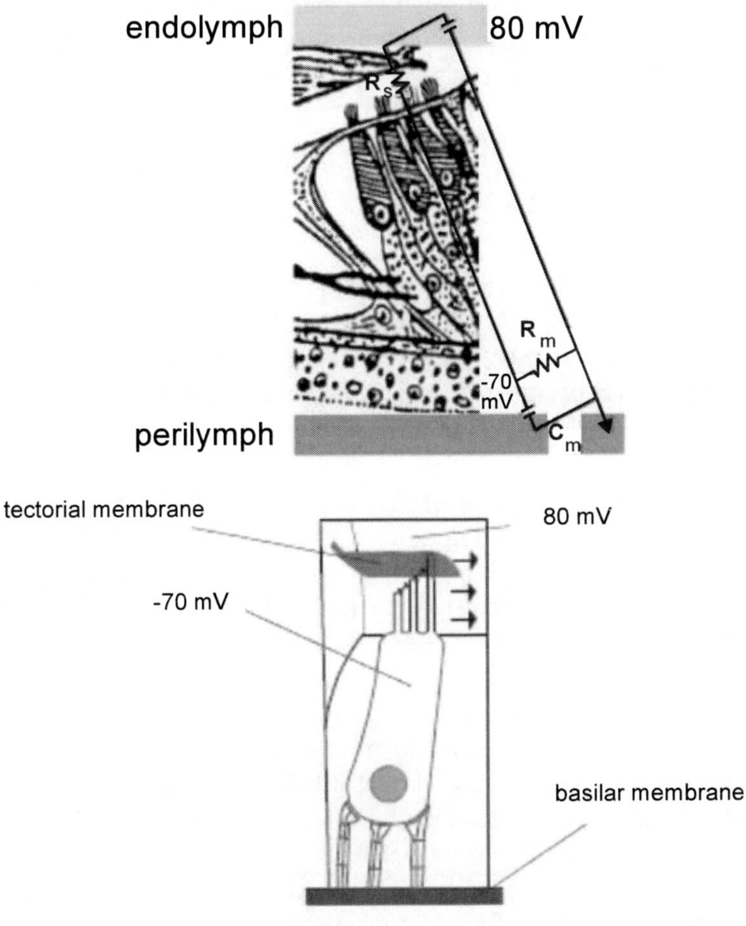

Figure 6. Equivalent electrical circuit and functioning of the Organ of Corti.

Since hair cells are connected to the basilar membrane - which is the upper limit of the scala tympani filled with perilymph – they are subjected to an electrical potential of -70mV. On the extern, the same cells are connected with the tectorial membrane, which is the lower limit of the scala vestibule filled with endolymph. This has an electrical potential of +80mV. The sum of the total electrical potential is equal to 150mV with electrical currents up to $8*10^{-9}$ A [13]. Signal transmission happens because of stereocilia deflection that produces a decrease in the electrical resistance R_s. This decrease produces an electrical potential drop across the resistance of the membrane R_m and an increase in the electrical potential in the hair cell itself. The phenomenon depends on frequency of the sound because this parameter changes the electrical capacitance of the membrane C_m [14].

The functions of the Organ of Corti are extremely special, and has only recently been fully elucidated with particular reference to the mechanism of amplification of the received acoustic signal [15]. Amplification comes from the driving action of the outer hair cells, which - as already said- are organized in triplets that amplify the oscillations of the cochlea canal from the movements of the endolymph. Through actions of expansion or contraction, they leverage on the Organ of Corti to generate forces capable of neutralizing the viscous resistance of the cochlear membrane, which in turn allows the elastic oscillations to be transmitted to the basilar membrane.

2.4. Sound Transmission to the Human Ear

The transmission mechanism, through which hearing organ sends a sound to the brain centers, is extremely complex [16]. The pinna collects sound waves and directs them to the auditory canal. When sound reaches the tympanic membrane, it causes the vibration of the tympani. Through the membrane, waves are transmitted to the middle ear without changes in frequency or amplitude. Then, vibration reaches the ossicles and is transmitted to the oval window. This time, unlike what happens at the interface with the outer ear, acoustic waves are modified. During transmissions to the inner ear, the anatomic structure of the organ allows an amplification of the sound pressure acting on the oval window. This amplification is nearly 100 times the amplification acting on the tympanic membrane outer wall. In addition, the vibration amplitude of transmitted sound waves is significantly reduced [13]. Pressure amplification happens for mainly four reasons:

a) Three bones act as a lever with an amplification factor of nearly 3. The location of the fulcrum is closer to the oval window than to the tympanic membrane. This causes an increase in the transmitted force in comparison to those acting on the membrane itself.

b) The surface on which pressure is applied is smaller. The oval window is on average 30 times smaller than the tympanic membrane (2.5 mm^2 and 80 mm^2 respectively).
c) The braincase acts as a sounding board.
d) The resonant effect on the acoustic meatus.

This complex system of sound transmission constitutes a kind of mechanical transformer. It works as a hydraulic press and increases the portion of absorbed energy at the expense of that reflected. Basically, it allows without great losses, the transmission of energy of the incident sound. The function of the stapedius muscle and tensor tympani enables the adaptation of the operation of the ear to the type of incident disturbance [14]. For instance, consider a sound having low intensity. When the wave hits the tympanic membrane, tensor tympani causes a refinement of the membrane. Because of the section reduction, the membane will vibe with more energy allowing a better auditory sensation. On the other hand, if the sound hitting the membrane is a high-intensity and quickly raising signal, stapedius muscle protects the ear from possible damage by reducing excursions of the ossicles and reducing the energy moving from the tympanum to the oval window. This ear defensive mechanism intervenes with a characteristic muscle reflex time (nearly 1 second). Because the sound waves come to the oval window before the defensive mechanism can be activated, it is not effective against impulsive noises [13]. Sound waves, transmitted through the previously described mechanism, arrive filtered to the oval window. Here the energy is transmitted to the incompressible fluid, the perilymph, which is placed in the scala vestibule. Vibration propagates along the scala and through the helicotrema, reaches the scala tympanic. At its end - closed by a membrane called secondary tympanic - there is the round window which allows the perilymph to oscillate. Figure 7 shows a stretched-out section of the cochlea.

For each frequency, there is a portion of the membrane characterized by maximum vibration amplitude. High frequency vibrations excite the areas close to the oval window (beginning of the cochlear duct), while low frequency vibrations excite those close to the helicotrema. For humans, the lower frequency of the audible range is 20Hz. Because of the length of the cochlear canal (30 millimeters) [10], sound pressure, which propagates on the scala vestibule, reaches the helicotrema in a time equal to a period. Therefore, no pressure difference is generated between the two scales and the basilar membrane is not excited. The 20Hz limit is due to the fact that if the ear were sensitive to lower frequencies, it would be possible to detect muscle contractions and impact with the ground as well. The upper limit of the audible range is, instead, 20kHz [17]. If the frequency of the sound exceeds this value, the mechanical inertia of the three bones does not allow the transmission of the sound wave to the perilymph, preventing the hearing organ from perceiving the sound.

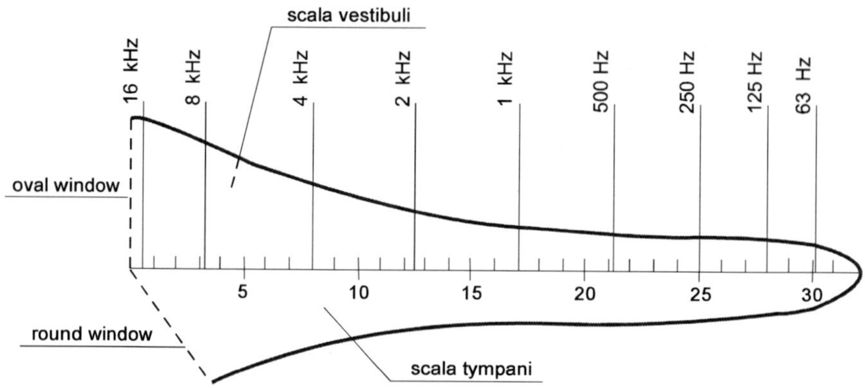

Figure 7. Stretched-out section of the cochlea.

To conclude, it is important to point out that the dissymmetry of the vibrating system produces a non-linear response when a sound wave has high intensity impacts on the tympanic membrane. Because of this dissymmetry, when a pure tone hits the hearing organ, the ear will perceive the harmonics of the incident sound. These are called subjective harmonics and are easily generated with low pitch sounds. When two pure tones with enough high intensity reaches the tympanic membrane, several different sounds are perceived. The two fundamental frequencies are the partial sounds in harmonic relation with the two pure tones, and the linear combination of the two fundamental tones. The fundamental tones are called combination tones and because of them, the brain can reconstruct a complex sound even if it does not get all the information about the original sound. Based on what is stated in this paragraph, it is clear that cochlea is not only an acoustic signal transducer for the brain, but it also performs functions of signal processing, favoring its decoding. In addition, the cochlea eliminates the weak signals at frequencies close to those characterized by greater intensity in order to suppress the noises that may disturb the main signal. This happens because the cochlear amplification system is saturated by the signals of higher intensity. Furthermore, cochlea promotes perfect decoding of complex sounds, which are characterized by tones that quickly change their amplitude and frequency. It transforms the signal so that tones with different intensities can produce an effect on the acoustic nerve stimuli of almost equal intensity.

3. Acoustic Descriptors and Characteristics of Sound Events

The human ear is clearly a uniquely sophisticated yet delicate organ. It can be compared to a sensing device able to detect change in acoustic pressure and vibration. This paragraph addresses the relevance between sounds and the above-described elements, introducing the main acoustic descriptors and how they can be used to evaluate possible risks arising from noise. In particular, phenomena of impulsive sounds and the change of sensitivity at the different frequencies is analyzed.

Usually, the acoustic descriptors used to characterize the acoustic phenomenon are the following [18]:

a) Effective A-frequency weighted SPLs (L_{AS}, L_{AF}, L_{AI} values determined using slow, fast, and impulse time constant respectively)
b) Maximum A-frequency weighted SPLs (L_{ASmax}, L_{AFmax}, L_{AImax} maximum values determined using slow, fast, and impulse time constant respectivel)
c) Equivalent continuous A-weighted sound pressure level $L_{A,\,eq}$-
   This is the constant noise level that would result in the same total sound energy (or the same mean square pressure) produced over a given period T by the real fluctuating noise. Basically, it is the stationary SPL that in the considered time T, produces the same effect of the real sound, whose level changes in time. The equivalent continuous A-weighted SPL can be evaluated using the following equation:

$$L_{A,eq} = 10\log\left[\frac{1}{T}\int_0^T\left(\frac{p\,(t)}{p_0}\right)^2 dt\right] \quad (1)$$

where T is the measurement time, p (t) the instantaneous value of the sound pressure in Pascal, and p_0 is the reference value of pressure (20 μPa). From a graphical point of view, the equivalent continuous A-weighted sound pressure level is the constant SPL that in the considered time subtends an area equivalent to that subtended by the real curve representing the evolution of the real phenomenon as shown in Figure 8.

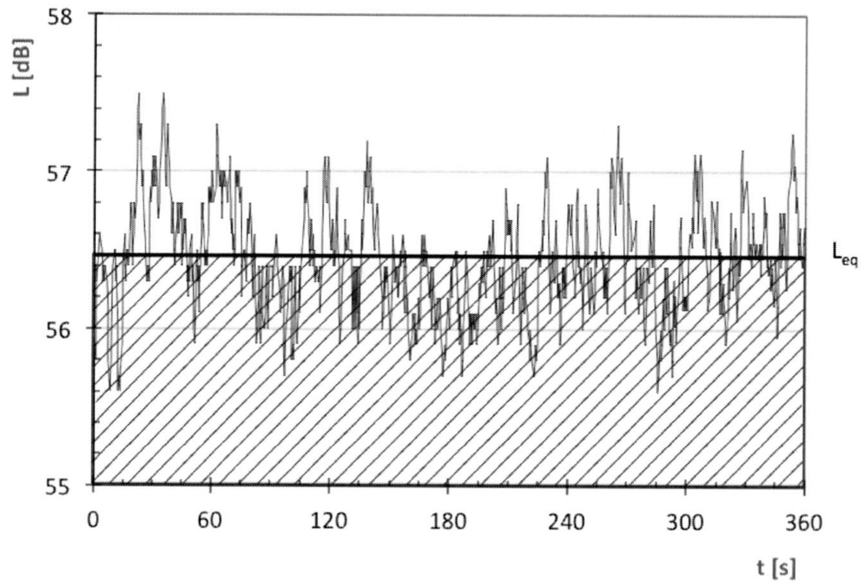

Figure 8. Graphical representation of the equivalent continuous SPL.

Historically, because of limited computational and analysis capacities, the continuous equivalent SPL has been the only parameter used. The choice of $L_{A,\ eq}$ is also determined from a link connecting it to hearing loss. This is carried out during an evaluation of effects on individuals who are subjected to constant SPL for a sufficiently long time (e. g.: workplaces) [19]. The equivalent continuous level allows the sound to be characterized with only the value of the sound energy produced by all sources present, regardless of their characteristics (fluctuating or continuous sources, impulsive, etc.). By contrast, it does not enable the identification of the individual noisy events that have contributed to the overall value. Therefore, it is clear that the $L_{A,\ eq}$ is not suitable to completely describe all the sound characteristics, especially for those phenomena which are extremely dynamic, or when data usable for identifying corrective operations is required.

d) A-weighted sound exposure level L_{AE};

This is the SPL of a sound that is constant for 1 second, which has the same sound energy of the measured real event. Practically, it is the SPL of a phenomenon that in 1 second produces the same energy produced by the phenomenon being studied in the period $T = t_2 - t_1$. It can also be seen as an equivalently continuous SPL normalized to 1 second. L_{AE} can be evaluated using the following equation:

$$L_{AE} = 10 \log\left[\frac{1}{T_0}\int_{t_1}^{t_2}\left(\frac{p(t)}{p_0}\right)^2 dt\right] \tag{2}$$

where t_2-t_1 is the time in which the phenomenon is measured and T_0 the reference duration (1 second). The relation existing between $L_{A,\,eq}$ and L_{AE} is shown here:

$$L_{AE} = L_{A,eq} + 10 \log\left(\frac{T}{T_0}\right) \tag{3}$$

3.1. Characterization of Impulsive Noise

Impulse noise is a category of noise which includes unwanted, almost instantaneous sharp sounds. Using the above-described parameters, it is possible to characterize the peculiarity of a sound. To detect if a sound can be classified as impulsive or not, the following parameters have to be measured and analyzed: L_{AFmax}, L_{ASmax}, L_{AImax}. According to current standards and legislative acts, a sound can be considered as impulsive when the following characteristics recur[20]:

- The difference between L_{AImax} and L_{ASmax} is bigger than 6 dB;
- The value of L_{AFmax} decays of at least 10 dB in less than 1 second;
- The event occurs at least 10 times during the day reference time and at least 2 times during the night reference time.

Figure 9 shows a practical example for the individuation of impulsive events. The graph represents the time histories of the phenomenon being studied, obtained using the L_{AFmax}, L_{ASmax}, L_{AImax} time constants.

Between the two vertical red lines (which represent a time span of 1 second from 09:36:05.700 to 09:36:06.700) the trend of the three profiles is studied. The difference between L_{AImax} and L_{ASmax} is bigger that 6 dB(A) (L_{AImax} - L_{ASmax} = 9.7 dB(A)) and the value of L_{AFmax} falls to 29.8 dB(A). In the above shown example, in 40 seconds seven impulsive events can be observed.

Over a longer time span (e.g.: 8 hours), these events are too short in time to increase the overall energy of the studied phenomenon and its continuous equivalent sound pressure level. Nevertheless, they are potentially harmful. They are characterized by extremely high-energy content and can probably reach the vestibular window before any defense mechanisms can be activated [21]. As already explained, ear protection - in the case of high-intensity and quickly raising signals - is entrusted to the stapedius muscle, whose contractions reduce ossicles excursion. Basically, this action results in a reduction of the sound energy that propagates from the tympanic membrane to the oval window. The guaranteed protection is nearly 10 dB at 1000 Hz, because the excursion reduction of the ossicular chain results in a decrease of nearly 60 μPa [18]. This ear defensive mechanism intervenes with a characteristic muscle reflex time (nearly 1 second) [22]. For this reason, it is not effective against impulsive noises, because the sound waves come to the oval window before the defensive mechanism can be activated. Indeed, it is important to point out that impulsive events are characterized by a persistent propagation time of a few milliseconds. Impulsive events are absolutely faster than the time human ears need to respond with a defensive mechanism that is comparable with the time of a muscle reflex.

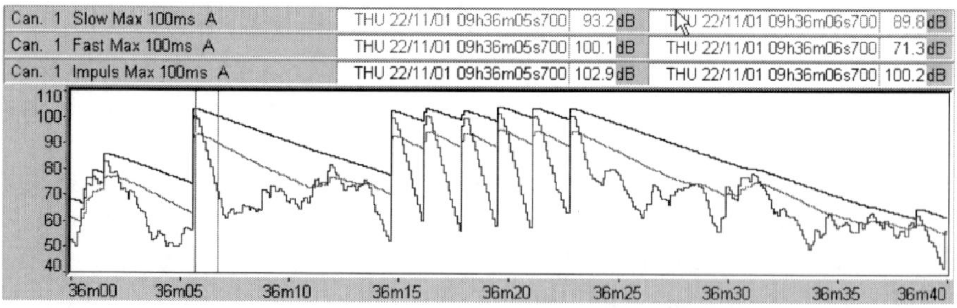

Figure 9. Time histories for the individuation of impulsive events.

3.2. Frequency Weight Curve

One of the main problems in the study of the psychoacoustic comes from the matching of the objectively measured acoustic parameters, with the subjective sensations they cause on the auditory system. The human's auditory response is neither constant with the frequency nor with the level [23]. A compensation of the levels at the different frequencies is required to obtain a suitable approximation of the human sensation. Frequency weighted curves are used in order to do so. Starting from the equal-loudness curve on the normal equal-loudness contours proceeded from Fletcher and Manson we can first obtain these curves and then standardize them to the ISO 226:2003 [24,25]. An equal-loudness contour is a measure of SPL over the frequency spectrum, for which a listener perceives a constant loudness when presented with pure steady tones. The unit of measurement for loudness levels is the phon, and is obtained through the reference to equal-loudness contours. By definition, two pure tones of differing frequencies are said to have equal-loudness level (i.e.: the same levels in phon) if they are perceived as equally loud.

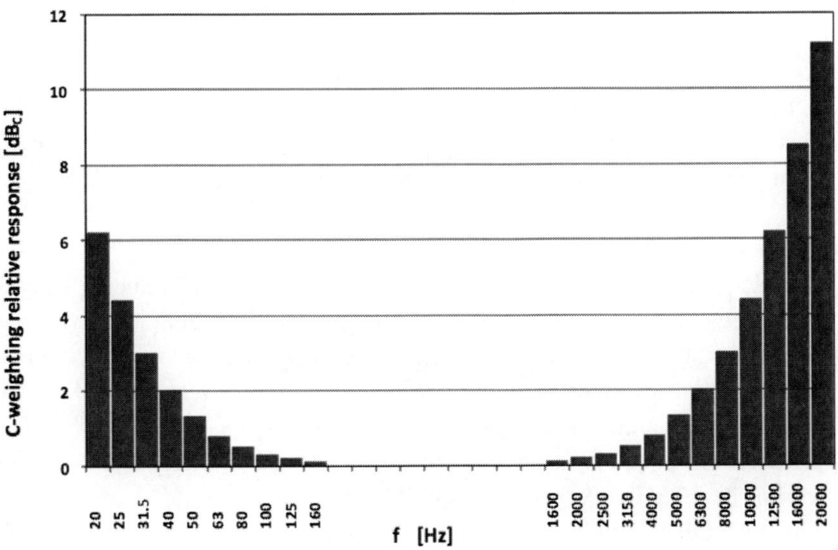

Figure 10. C-frequency weighted relative response.

Usually, four different frequency-weighted curves are individuated. The A-frequency weighted curve is the inverse of the 40-phon equal-loudness contour. It is the most commonly used and approximates the behavior of the human ear in the low-frequencies range. The B-frequency weighted curve is the inverse of the 60-phon equal-loudness contour, but it is no longer used. The C-frequency weighted curve is the inverse of the 100-phon equal-loudness contour, while the D-frequency weighted curve is the inverse of the 40-noy curve. Basically, through the use of these curves, the energetic content of the acoustic signal is modified (decreased or increased) in as a function of its frequencies. As shown in Figure 10, when a signal is processed using the C-frequency weighted curve it is artificially lowered in low frequencies up to 160 Hz and from 1600 Hz on. As a result, the total energy of the also signal decreases.

4. Workers Protection from Noise and Normative Scenario

Workers protection from the risks arising from prolonged exposure to a noisy processes is one of the trickiest pollutant issues to solve. The problem is made even more complicated as the possible solutions of an event that produces potentially harmful contexts to the physical integrity of individuals cannot always be implemented. For instance, even use of headsets is often prohibited as they may induce unacceptable limitations for other aspects of the workers' security. The analysis of acoustic scenarios that may be found in different working activities is usually complex and the conclusions may not be free of errors because of the particular environmental conditions in which analysis is carried out.

In order to avoid conflicting analyses on the measured phenomena, several directives regarding workers' health-care and protection from noise have been issued. For instance, in the United States procedures for guaranteeing security of workers in the workplace are regulated by the OSHA (Occupational Safety & Health Administration) Standard "*1910 Occupational Safety and Health Standards, Subpart G*" in observance of the 1970 Law on safety and health in the workplace, (OSHA Act) [26].

On the other hand, in Europe, protection from noise in a work environment is upheld by *Directive 2003/10/EC of the European Parliament and of the Council of February 6th, 2003 on the minimum health and safety requirements regarding the exposure of workers to the risks arising from physical agents (noise)* [3]. The Directive supplies the minimum safety prescriptions to ensure health-care and protection from noise to those workers who are, or are likely to be exposed to risks from noise as a result of their work. The Directive also suggests priority to reducing risk at the source, to reduce the level of exposure to noise by incorporating preventive measures in the design of workstations and to select suitable personal protective equipment (PPE). Furthermore, it promotes provisions relating to work equipment and methods, thus contributing to the protection of the workers involved. All member states must conform their national directives according to what is stated in this European Directive.

The European standard uses the following parameters, calculated with a reference pressure of 20 μPa, for the protection of workers [3]:

a) Daily noise exposure level ($L_{EX,\ 8h}$);
b) Weekly noise exposure level ($L_{WE,\ 8h}$);
c) Peak sound pressure (p_{peak}).

The daily noise exposure level $L_{EX,\ 8h}$ is the time-weighted average of noise exposure levels for a nominal eight-hour working day as defined by international standard ISO 1999: 1990 [27], point 3 paragraph 6 (and its modifications and updates [28]), covers all those noises present in a workplace, including impulsive noise. On the other hand, the weekly noise exposure level $L_{WE,\ 8h}$ is the time-weighted average of noise exposure levels for a nominal week of five eight-hour working days, as defined again by international standard ISO 1999:1990, point 3 paragraph 6 and its modification and updates [27], [28]. Both are used to determine workers' exposure to noise during their working activities and they offer an averaged absorbed dose during their working hours. Additionally, the peak sound pressure $p_{peak,\ C}$ represents the maximum value of the instantaneous C-frequency weighted noise pressure. In reference to the pressure of 20 μPa, the Directive establishes the peak sound pressure maximum admissible value (exposure limit value) at $p_{peak,\ C}$ = 200 Pa [140 dB (C)]. Moreover, it states that the maximum value for which the use of individual hearing protection devices is not obligatory (lower exposure action values) is $p_{peak,\ C}$ = 112 Pa [135 dB (C)], while the maximum value for which usage of individual hearing protection devices is obligatory (upper exposure action value) is $p_{peak,\ C}$ = 140 Pa [137 dB (C)] [3]. This means that in those working environments where $L_{EX,\ 8h}$ is below 80 dB (A) (daily noise exposure level limit) and the $p_{peak,\ C}$ levels are less than 135 dB (C), the employer can refuse to supply workers with headsets and other personal protective equipment.

Contrary to what is stated in the earlier laws, the Directive currently requires the determination of peak sound pressure levels to be carried out using only frequency-weighting scales. Furthermore, no record of any other physical-quantity temporal evolution such as impulsive events is required, even if those events may be relevant for detecting the effective solicitations on the ear.

In summary, the Directive 2003/10/EC provides that analyses only have to be carried out by sampling the maximum C-frequency weighted value of the instantaneous sound pressure level $p_{peak,\ C}$ and the A-frequency weighted equivalent level normalized to a standardized 8 hours working day $L_{EX,\ 8,}$or to a standardized 40 hours working week $L_{WE,\ 8h}$. In these equivalent levels, all the noises produced in the working places include the impulsive ones. It is clear that the Directive lacks complete worker protection. Indeed, one of the main assumptions of this procedure is that impulsive events simply contribute to the determination of an exposure value and further, that their evaluation converges in the calculation of the previously mentioned action values ($L_{EX,\ 8h}$, $L_{WE,\ 8h}$). They are not separately considered as dangerous effects on the acoustic phenomenon. According to the Directive, for environments in which noise exposure levels are below the threshold, a simple evaluation of the $p_{peak,\ C}$ value is sufficient. No importance is given to signal's time history, which is the sampled-signal temporal trend.

While cases in which peak sound pressure levels are high (but still lower than the exposure action value) and noise exposure levels are below the action levels may seem rare, they are extremely common situations in workplaces characterized by occasional loud impulsive sounds but a short temporal persistence. For instance: material stress testing laboratories, mechanical workshop activities, metal-plates pressing operations, furnaces' crucible cleaning after the production of ferrous materials, etc. Other limitations of the Directive are embodied in the choice of the C-frequency weighting curve, which corresponds to the inverse of 100-phon equal-loudness curve on the normal equal-loudness-level chart [25]. This curve artificially decreases the value of the SPLs at low and high frequencies. Therefore, it may lead to an erroneous evaluation of peak value and therefore to an inadequate protection of the workers.

As stated before, the Directive fixes the value of $p_{peak,\ C}$ = 112 Pa [135 dB (C)] as the maximum admissible value for which any individual hearing protection devices is not obligatory. It is possible to highlight how this setting is actually unreasonable. Indeed, analyses of the normalized audiogram [25] of the human ear show how the threshold of pain is less than 135 dB for all those frequencies in the audibility range (1 – 8 kHz). Moreover, for frequencies above 1 kHz this threshold is much lower than the value chosen as a reference in the norm. In addition, several medical diagnoses and studies have demonstrated that repeated exposures for middle-long periods to impulsive noise even significantly lower than 135 dB, can produce irreversible damage to the ear [29].

Negative implications embodied in this procedure of noise measurement, along with the assumption that impulsive events contribute solely to the determination of the entity of exposition only, are clearly an excessive simplification that may lead to shortcomings in workers protection. This is because the potentially harmful events are omitted in the analyses and the signal is artificially depowered. From the experience gained with the analyses carried out, it is evident that acoustic-characteristics of a work place are poorly represented when the descriptors indicated in the Directive (p_{peak}, $L_{EX, 8h}$, $L_{WE, 8h}$) are used. As a result, several situations of risk, which in time may cause irreversible damage to workers, are considered non-hazardous and therefore the procedures for protecting workers health are not implemented at all. Indeed, the main failures of the investigation pattern can be summarized as following:

a) Inability to identify, solely through the data for the peak sound pressure ppeak, the incidence of energy on the human ear and to determine risky situations.
b) Inability to identify, solely through the data for the noise exposure levels $L_{EX, 8h}$ and $L_{WE, 8h}$, the incidence of energy on the human ear.
c) Inadequacy of weighting scale used for determining actual exposure to the maximum peak sound pressure.

For these reasons, the authors have individuated an integration procedure mainly consisting of: the individuation of the time history describing the phenomenon being tested, the research of occurring impulsive events, the use of the un-weighted peak sound pressure levels and the study of the spectral composition of the analyzed sound with the possibility of using a Fast Fourier Transform (FFT) analysis.

5. Critical Analysis of Parameters Used in the Directive for Hearing Risk Determination

The incompleteness of the pattern currently used has emerged during numerous noise analyses that the Laboratorio di TCA at Universitá della Calabria Italy has carried out during its research and consulting work. Measures carried out in the last fifteen years have developed a large database of surveys regarding several working activities in different fields of the industrial sector (testing laboratories, repair shops, joineries, metalworking industries, etc.) under commission to the University by health authorities and courts. These surveys were conducted in response to a demand for medical specialists and magistrates who wished to gain knowledge about sound field peculiarities during working activities. These surveys were used to determine the possible causes of pathological phenomena onset in workers and whether or not these problems could be connected to the tasks carried out during work.

Critical analysis of sampled data clearly shows that descriptors used in the *Directive 2003/10/EC* do not always highlight the criticalities of work situations. Through all the completed analyses, the researchers have acquired data according to what is stated in the European and Italian regulations regarding acoustics signal surveys [20], [30] and have analyzed them according to prescriptions provided by European Community legislation [3], [31]. In some analyzed scenarios, descriptors indicated in the European Directive are sufficient to fully represent the real situation of risk and can be used to identify protection procedures for workers. Nevertheless, situations in which they can be used are related to scenarios not characterized by harsh acoustic environments.

In the following figures, measures related to activities for which it is possible to use noise exposure levels for evaluating the risk are reported. Figure 11, shows the time histories of the equivalent continuous A-weighted sound pressure level $L_{A,\,eq}$, the un-weighted peak sound pressure level p_{peak}, and other main acoustic descriptors (A-weighted single event level L_{AE}, A-weighted daily noise exposure level on 5.5 hours $L_{AEX,\,5.5}$, A-weighted daily noise exposure level normalized to 8 hours $L_{AEX,\,8h}$ and percentile levels L_{max}, L_{10}, L_{30}, L_{50}, L_{70}, L_{90}, L_{min}) calculated from rough data for activities carried out in nearly a half-day of work (5.5 hours), by an operator of a cast iron production factory. The instrumentation used consisted of a dosimeter, which is a device used to detect an individual or object exposure to noise. During this study, the Larson Davis 705 model was used. This device measures in the dynamic range 70 – 140 dB. In addition, it offers the possibility of storing the equivalent continuous sound pressure level L_{eq}, the un-weighted peak sound pressure level p_{peak} the A-weighted single event level L_{AE} and the percentile levels L_n in three different frequency-weighted scales: A, C and un-weighted, according to the ANSI S1.4 1983 norm [21]. The device was placed with a microphone on the shoulder of the subject being tested to detect the effective SPLs acting on his ear. The signal was sampled using the previously mentioned European and Italian laws and then analyzed focusing attention on the temporal evolution of the phenomenon.

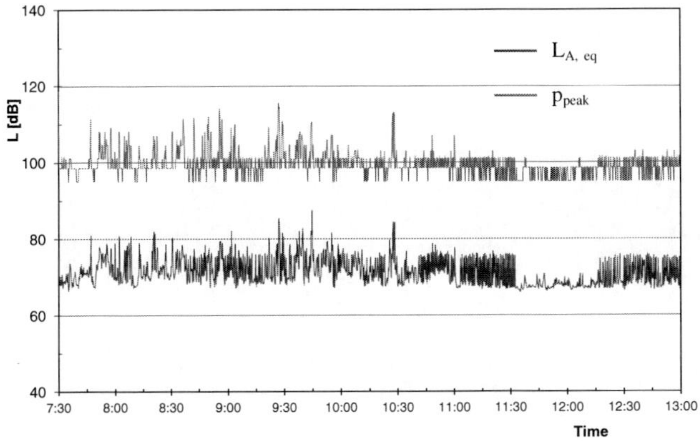

L_{AE}	112.0
$L_{AEX, 8h}$	71.5
$L_{AEX, 5.5h}$	73.1
L_{max}	107.2
L_{min}	65.4

L_{10}	75.7
L_{30}	71.1
L_{50}	69.2
L_{70}	68.0
L_{90}	67.1
P_{peak}	118.6

Figure 11. Time history measurements of the peak and equivalent continuous A-weighted sound pressure level for an industrial activity.

It is possible to observe that carrying out an analysis with procedures from the Directive 2003/10/EC is correct and allows for the identification of hazardous conditions for workers. In particular, it is possible to notice that the dose adsorbed by the worker during the cleaning activity has an A-weighted daily noise exposure level of 5.5 hours $L_{AEX, 5.5h}$ of 73.1 dB(A), which scaled on the whole working day correspond to an A-weighted sound pressure level normalized to 8 hours, $L_{AEX, 8h}$ equal to 71.5 dB(A). Both values are far below the threshold of 80 dB(A) indicated in the Directive, therefore, the only parameter that could force the worker to use a suitable personal protective equipment (PPE) is the C-frequency weighted peak sound pressure value $p_{peak, C}$. In the considered example, the un-weighted peak sound pressure value p_{peak} is equal to 118.6 dB, which evaluated using the C-frequency weighted curve is far below the threshold of 135 dB(C) which is the lower exposure action values presented in the Directive.

Cleaning of a crucible is a quite noisy activity. If the Directive can ensure an effective protection in this situation, this could be legitimate thinking that can ensure the same protection even in situation less noisy. Nevertheless, the following example shows that the current legislation has several weaknesses, which render protection not as effective as its original aims. This can be highlighted analyzing the SPLs of Figure 12 and Figure 13, which show a dosimetric analysis relating to activities carried out in 6 hours by an operator in an open-to-the-public secretarial office.

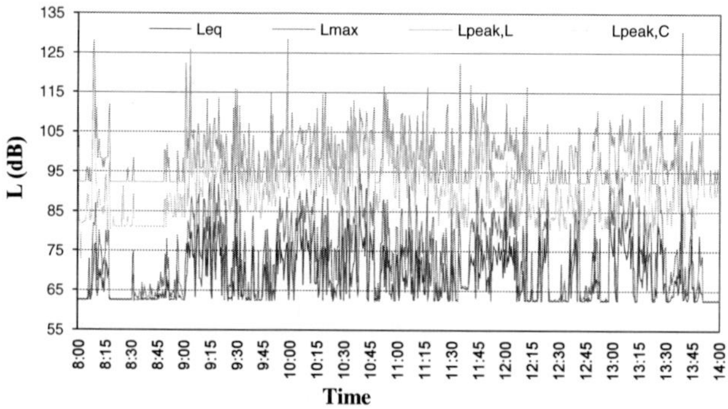

L_{AE}	111.5
$L_{AEX, 8h}$	51.1
$L_{AEX, 6h}$	68.1
L_{max}	104.8
L_{min}	62.6

L_{10}	77.2
L_{30}	75.8
L_{50}	66.5
L_{70}	62.9
L_{90}	62.5
$P_{peak,C}$	118.2
$P_{peak, L}$	130.7

Figure 12. $L_{A, eq}$, L_{max}, $L_{peak, L}$, and $L_{peak, C}$ time histories for an office work.

This activity is clearly less noisy than that described in the previous example. It as can also be observed from the value of the A-weighted sound pressure level normalized to 8 hours, $L_{AEX, 8h}$ equal to 51.1 dB(A), nearly 20 dB(A) smaller than the $L_{AEX, 8h}$ which characterizes the industrial activity. The value of the daily noise exposure level is smaller than the limit presented in the Directive, as well as the value of the peak sound pressure [130.7 dB]. As stated before, the Directive fixes the value of $p_{peak, C}$ = 135 dB (C) as the maximum admissible value for which use of any individual hearing protection devices is not obligatory. Again, it is worth noting that the value of the acoustic peak presented in Figure 12 is still un-weighted, so when the C-frequency weighted curve is applied, maximum value will decrease and it will be lower. Nevertheless, it is possible to highlight how the setting stated in the Directive is actually unreasonable When data is analyzed using prescriptions presented in the Directive, the working place being studied is safe for workers' health. Form Figure 13 one can observe that this is not completely true because at least four events having intensity bigger than the threshold of pain occur.

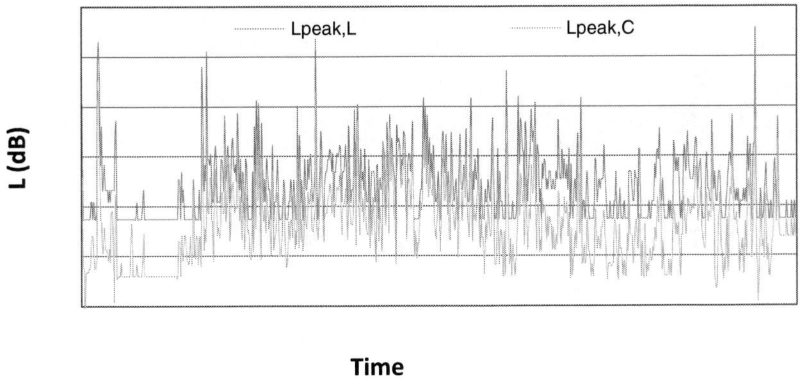

Figure 13. Comparison between $L_{peak,\ L}$ and $L_{peak,\ C}$ time history for an office work.

Another interesting example that shows how the currently used regulation does not allow for the presence of risks to the organ of hearing, is the noise to which public urban transportation drivers are exposed. Several surveys were carried out during the working hours of public transportation drivers in the city of Cosenza, Italy. Three different buses were used for the analyses: an IRISBUS 203e 9.27/CNG, a BREDA MENARINI BUS M24, and an IVECO 49e. The First two are methane-fueled vehicles which are used on the "Circolare veloce" route. This line connects the Northern bound with Southern bound of the city with a low penetration (about 30% of the whole route) in the urban fabric. The IRISBUS 203e 9.27/CNG has a length of 9.00 meters, a width of 2.35 meters, and weighs 9000 kilograms. The BREDA MENARINI BUS M24 has a length of 12.10 meters, a width of 2.55 meters, and weighs 11200 kilograms. The third one, instead, is a small gasoline-fueled vehicle used to connect the historical part of the city with the valley area and has a length of 8.10 meters and a width of 2.00 meters.

Drivers were supplied with two Larson Davis 705 dosimeters; one set to detect the C-frequency weighted peak sound pressure value $p_{peak,\ C}$, the other set to measure the un-weighted peak sound pressure value p_{peak}.

Figure 14 shows the SPL time history recorded on the smallest bus, the IVECO 49e during the seven hours work shift of the driver, while Figure 15 shows the detail of the C-frequency weighted peak sound pressure level and the un- weighted peak sound pressure level.

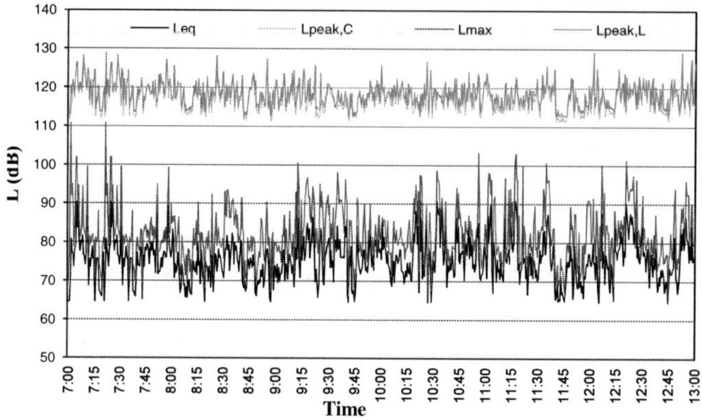

L_{AE}	107.4
$L_{AEX,\ 8h}$	52.8
$L_{AEX,\ 6h}$	69.6
L_{max}	101.5
L_{min}	62.3

L_{10}	75.7
L_{30}	75.8
L_{50}	66.5
L_{70}	62.9
L_{90}	62.5
$P_{peak,\ C}$	118.6
$P_{peak,\ L}$	119.0

Figure 14. $L_{A,\ eq}$, L_{max}, $L_{peak,\ L}$, and $L_{peak,\ C}$ time histories for the driver of the IVECO 49e bus.

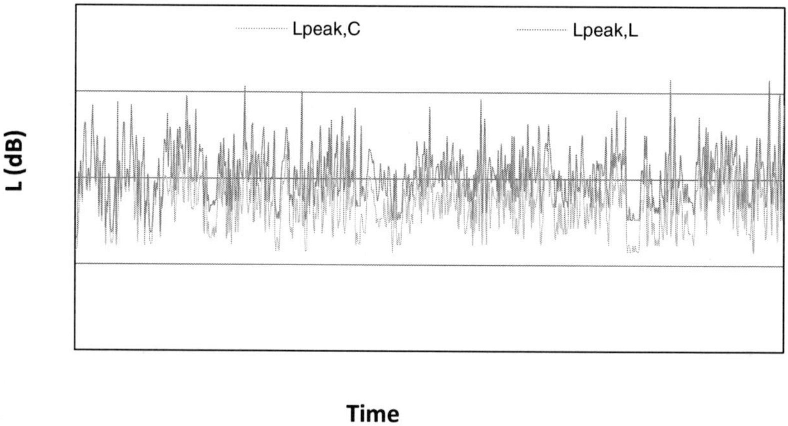

Figure 15. Comparison between $L_{peak,\ L}$ and $L_{peak,\ C}$ time history for the driver of the IVECO 49e bus.

It is possible to observe that the daily exposure dose is smaller than the value prescribed in the Directive. Therefore, the only analysis that can be done is with regards to the value of the peak sound pressure. In this case, both the C-frequency weighted and the un-weighted values are almost the same below the threshold indicated in the legislation and below the threshold of pain. This happens because of the diesel engine that propels the vehicle. The diesel engine's acoustic emissions are mainly in the middle range of the audible range (200 – 1500 Hz), and therefore, are not influenced by the penalization of the C-frequency weight curve. Since the IVECO49e bus goes through urban streets not affected by high volumes of traffic and it moves at low speed, the use of Directive 2003/10/CE does not imply errors in the evaluation of the noise to which workers are subjected. The situation is different if data collected on the methane-fueled buses are analyzed. The following figures 16, through 19 shown the time histories measured on the IRISBUS 203e 9.27/CNG and BREDA MENARINI BUS M24 busses.

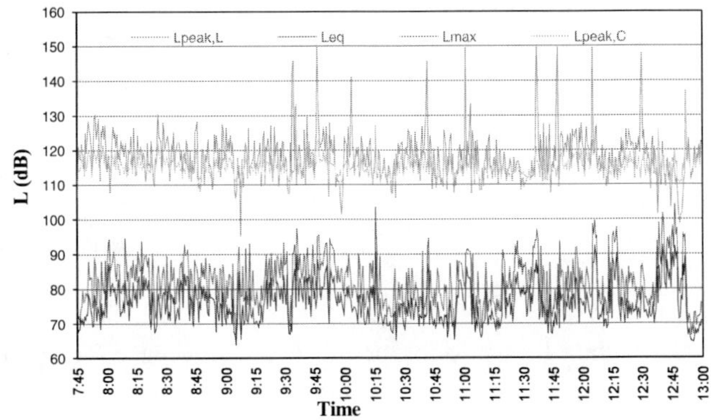

L_{AE}	120.6
$L_{AEX,\ 8h}$	51.1
$L_{AEX,\ 6.25h}$	77.8
L_{max}	104.2
L_{min}	64.3

L_{10}	87.6
L_{30}	84.7
L_{50}	76.9
L_{70}	73.3
L_{90}	70.7
$P_{peak,\ C}$	133.4
$P_{peak,\ L}$	149.6

Figure 16. $L_{A,\ eq}$, L_{max}, $L_{peak,\ L}$, and $L_{peak,\ C}$ time histories for the driver of the IRISBUS 203e 9.27/CNG bus.

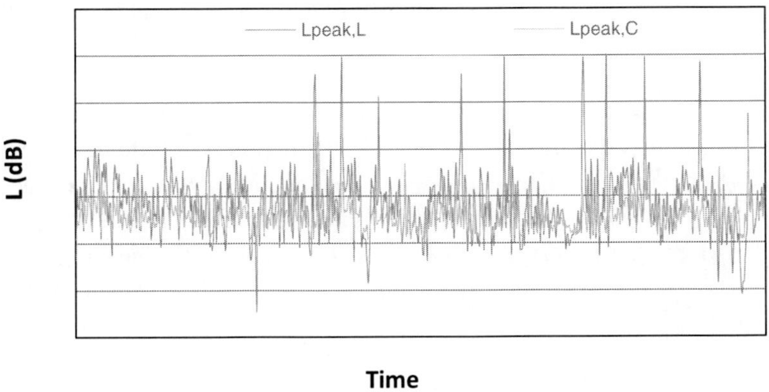

Figure 17. Comparison between $L_{peak,\ L}$ and $L_{peak,\ C}$ time history for the driver of the IRISBUS 203e 9.27/CNG bus.

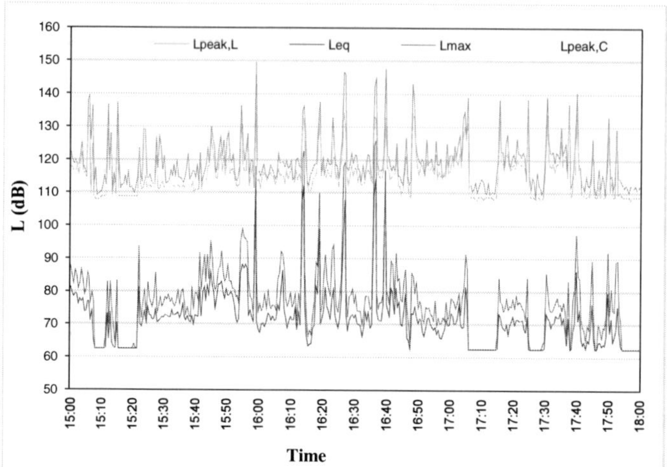

L_{AE}	113.1
$L_{AEX,\ 8h}$	27.3
$L_{AEX,\ 3h}$	72.7
L_{max}	125.8
L_{min}	62.6

L_{10}	79.9
L_{30}	78.1
L_{50}	72.7
L_{70}	70.1
L_{90}	66.5
$P_{peak,\ C}$	134.8
$P_{peak,\ L}$	149.6

Figure 18. $L_{A,\ eq}$, L_{max}, $L_{peak,\ L}$, and $L_{peak,\ C}$ histories for the driver of the BREDA MENARINI BUS M24 bus.

Through the analysis of data it is possible to observe how the equivalent continuous A-weighted sound pressure level is below the limit expressed in the Directive. It is equal to 77.8 dB(A) and 72.7 dB(A) respectively, and becomes even smaller when it is scaled and normalized to 8 hours. One more time, the most important observation can be done on the peak sound pressure. In the first considered example, the un-weighted peak sound pressure value p_{peak} is equal to 149.6 dB, which evaluated using the C-frequency weighted curve becomes $p_{peak,\ C}$ = 134.3 dB (C). One can observe that when the un-weighted conditions are considered, the maximum peak value is even higher than the upper exposure action value (value for which usage of individual hearing protection devices is obligatory). Nevertheless, when the C-frequency weighted curve is applied it becomes smaller than the lower exposure action values [135 dB (C)] presented in the legislation and therefore the use of individual hearing protection devices is not obligatory. In this case, since values for daily exposure and peak sound pressure are both smaller than the lower value of action, there is no obligation for the employer to provide workers with individual hearing protecting devices, even if they are exposed to sounds potentially harmful to the ear.

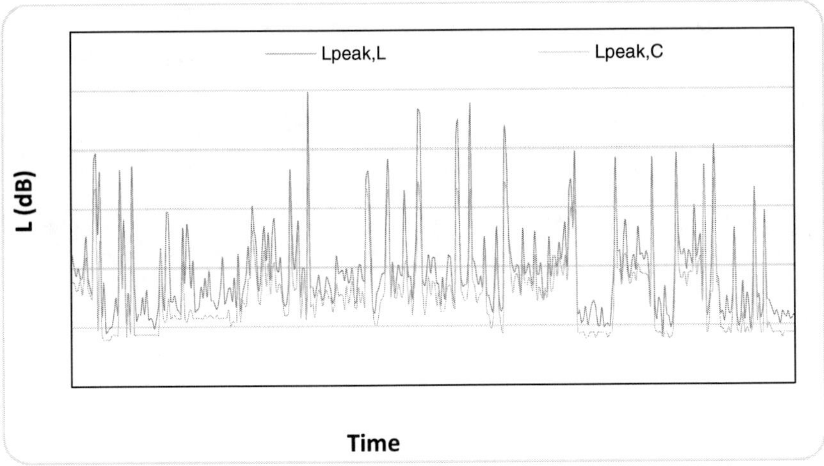

Figure 19. Comparison between $L_{peak, L}$ and $L_{peak, C}$ time history for the driver of the BREDA MENARINI BUS M24 bus.

On the other hand, it is important to note that if the value of the parameter adopted in the previously used Directive 86/188/CEE [33] (which refers to the un-weighted value) was still stored, the worker would have to be supplied with PPEs. Indeed, one of the considerations the authors would suggest - at least concerning peak sound pressure value - is to refer to the limit reported in that Directive. The same consideration can be done observing data recorded on the BREDA MENARINI BUS M24 bus, where the same values are reached. It is important to note that a sound having intensity close to 150 dB may cause irreversible damages to the laceration of the tympanic membrane. In addition to these damages, the problem is made more complex by the presence of impulsive components, which are completely ignored in the Directive 2003/10/EC.

5.1. Determination of the Energy Discharged on the Ear Using P_{peak}, $L_{EX, 8h}$, $L_{WE, 8h}$

The problem of the impulsive events can be highlighted when the following surveys are considered. Figure 20 shows in detail the temporal profiles of the un-weighted peak level and maximum impulse level, recorded with a sampling time of 60 milliseconds in a mechanical workshop.

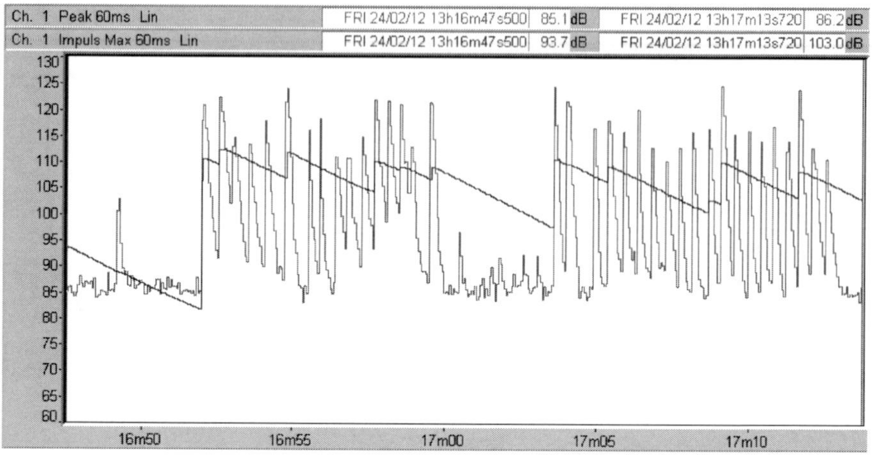

Figure 20. Detail of the linear peak and maximum impulsive level time histories recorded in a mechanical workshop.

A comparative analysis shows that peak levels (in red) change faster than impulsive signal (in blue), which represents an input capable of stressing the ear. In addition, the presence of several components having high energy is completely ignored for risk determination purposes. Indeed, acquisition with the Impulse time constant remains inhibited for all the time in which the pulse signal decreases. This happens because the tool is not able to record other signals before it is completely discharged. Acquisition of the maximum value of p_{peak} can insure a protection only if attention or risk levels are reached, but it does not consider all the energy that is discharged on the tympanic membrane. Basically, stapedius defensive mechanism is not activated and damages may happen in the event of prolonged exposures.

When numerical data is considered, is it possible to note that if the threshold regarding the daily exposure dose $L_{EX,\ 8h}$ is not reached (because when the 30 minutes signal is normalized to 8 hours, the equivalent level will become smaller), the only evaluation that can be done is about the value of the single peak sound pressure p_{peak}. In the example shown this value is equal to 125.6 dB, which is smaller than the value pointed out in the Directive. On the other hand, several energetic events having level bigger that 120 dB happen. The Directive does not consider them at all, even if they are potentially dangerous. Indeed, several studies have demonstrated how damages of impulsive noise on hearing loss are much more than that of continuous and stationary noise [32]. Because of stationary sounds, the ear can adopt constant and steady state protection mechanisms until it reduces the intensity of the pressure acting on ear's inner parts [13]. However, for variable and impulsive sounds the ear cannot apply the same protections. This is a clear example of a simple evaluation of the exposure limit values, which is not enough to ensure workers protection.

5.2. Importance of Impulsive Events on the Ear

Experimental data collected over a long period of time shows that analysis procedures contained in the Directive does not seem to be appropriate to protect the health of workers for all cases that may occur, especially when events that repeat over short time period are considered. Determination of the effects is unsatisfactory because these events do not have enough energy for reaching the threshold indicated in the Directive. However, they are potentially harmful since they can reach the vestibular window before defensive mechanism may be activated [21].

The stapedius muscle, which contraction reduces the oscillation of the ossicles and limits the acoustic energy transmitted through the tympanic membrane, guarantees principal protection from these events. Defensive mechanisms happen with the typical time characteristic of a muscle reflex. Therefore, it is not effective against sounds characterized by a persistence of a few milliseconds. Effects of impulsive events are extremely dangerous, bot for the particular conformation of the ear and the mechanism of stimuli transmission to the brain.

Impulsive events occur more frequently than thought. Sometimes they are characterized by low SPLs, but at the same time by a repetitiveness that is likely to go unnoticed. This is a significant source of danger due to the accumulation of energy on the tympanic membrane. The stress condition can be extremely serious because other factors may aggravate the situation. Very often the particular spectral energy distribution amplifies the problems. An example of this situation can be shown using the data collected on the buses for urban transport in the city of Cosenza.

Beyond the two dosimeters placed close to drivers' ear a Class 1 real time SPL Meter "Symphony" by 01dB with preamplifier and microphone from the same manufacturer was used. It recorded the SPLs in the cockpit indicating points from the Italian Directive for the measurement of environmental noise [20]. Figure 21 shows the A-frequency weighted SPL time history obtained using the Slow time constant and recorded during the journey from the city center to Northern terminus.

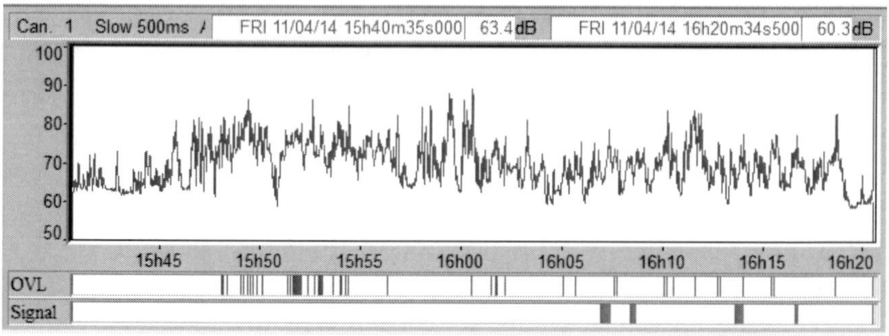

Figure 21. A-frequency weighted SPL time history recorded in the bus cockpit.

Figure 22 shows the same journey, and for a time lag including within the previous, the SPL trend obtained using the time constant Slow (red line), Fast (blue) and Impulse (grey). It is clear the presence of several components having high energy. This characteristic is constant for the whole journey (not shown here for the sake of brevity) and it gives an idea of how important the energy overload on the driver's ear is.

In addition, in order to prove that situation is made worst because of the frequencies of the sound, an analysis in frequency domain is carried out. This consists of a FFT analyses using normalized 1/3 of octave band. Results show that energy is mainly released at low frequencies and that the most energetic components are below 50 Hz. This situation makes noise exposure particularly critical. Following figures, 23 and 24 show the frequency domain results when two different weighting contours are applied to the signal.

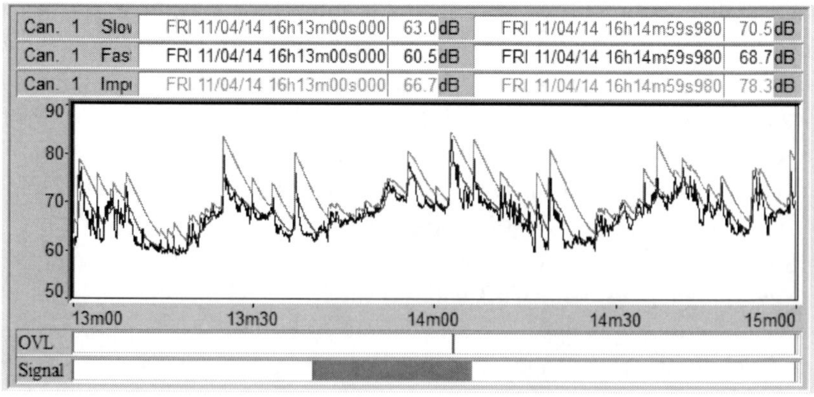

Figure 22. Detail of the A-frequency weighted SPL time history recorded in the bus cockpit with the S, F, I time constant.

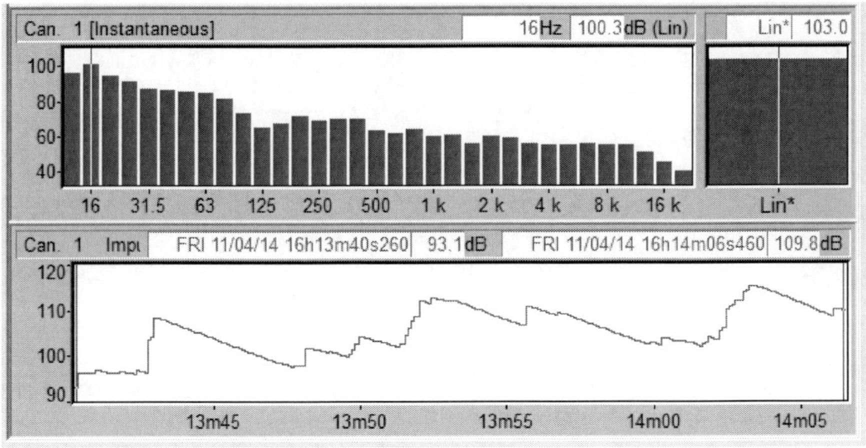

Figure 23. Un-weighted spectral composition (1/3 octave bands) for the signal recorded in the cockpit.

When the C-frequency weighted curve is used, it is possible to observe how the energy of the signal is dangerously underestimated. The situation may get worse when more energetic exposure levels are considered. The situation is particularly significant because during the journey, the stresses having a high value of the measured energy are numerous. The phenomenon varies rapidly in time, reaching the conditions which according to the rule are recognized as impulsive events.

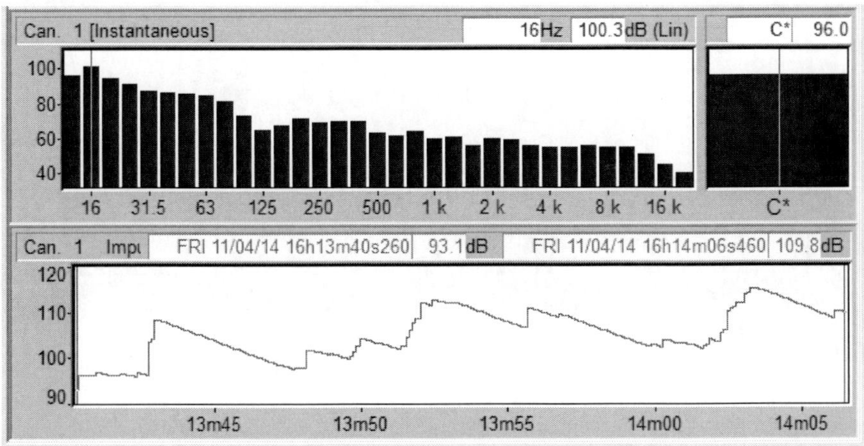

Figure 24. C-frequency weighted spectral composition (1/3 octave bands) for the signal recorded in the cockpit.

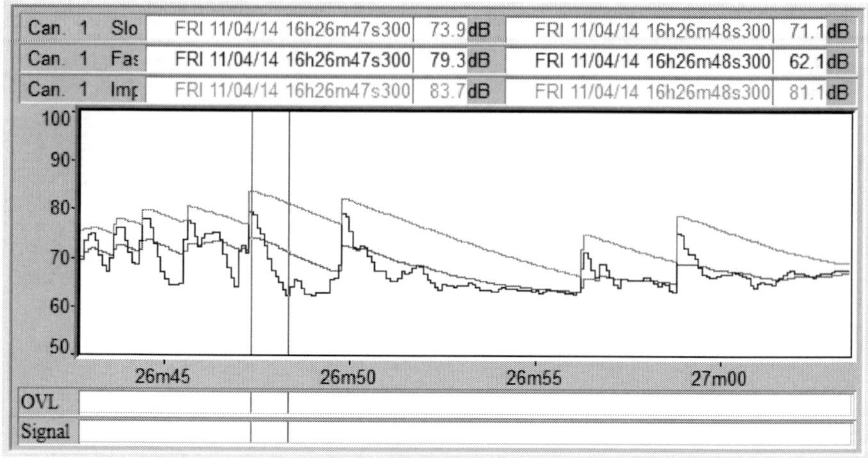

Figure 25. Impulsive event recorded at in the bus cockpit at 16.26.47.300.

Examples also demonstrate how these situations occur almost continuously. The solution to the specific problem is not simple, because drivers of a bus cannot protect themselves by using headphones insulators, which means that the problem must be solved more generally, in terms of the design of the means of transport.

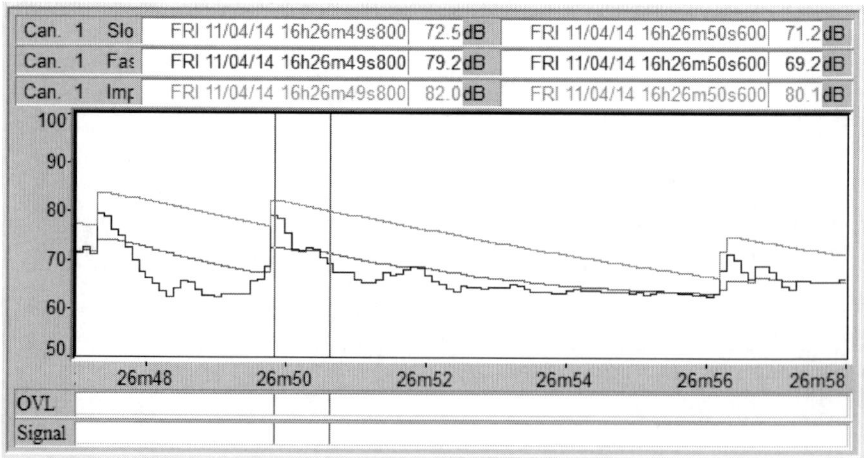

Figure 26. Impulsive event recorded at in the bus cockpit at 16.26.49.800.

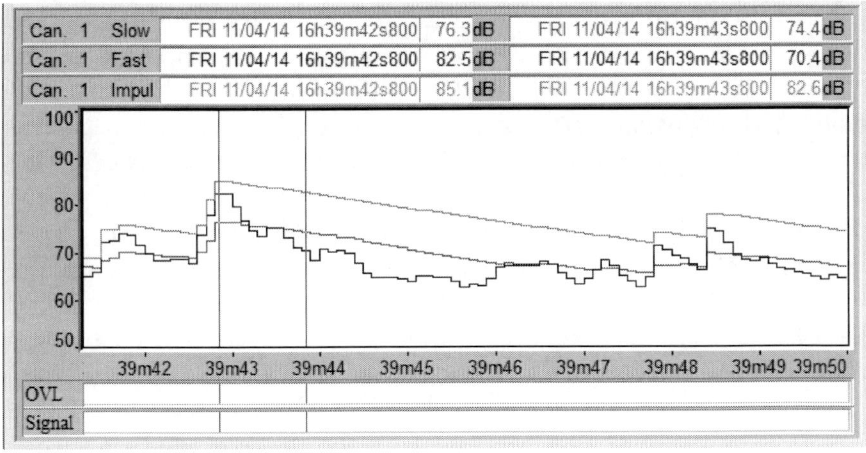

Figure 27. Impulsive event recorded at in the bus cockpit at 16.39.42.800.

Conclusion

Analyses regarding the minimum health and safety requirements for the exposure of workers to the risks arising from noise have been conducted in Europe since the early 1990s. Through years, several progresses have been achieved in the protection of workers. Nevertheless, the currently used legislation is not so effective as in its original aims. In this chapter, starting from the anatomic description of the characteristics of the hearing organ a description of the weaknesses embodied in the Directive 2003/10/EC of the European Parliament and of the Council of February 6th, 2003 on the minimum health and safety requirements regarding the exposure of workers to the risks arising from physical agents (noise) is carried out. The main acoustic descriptors are then presented along with the characteristic of the impulsive noise and the different frequency weighed contours.

As it has been observed in the first example provided, the Directive can individuate the most common critical situations, but in other instances it fails in guaranteeing effective protection of the most sensitive parts of the ear such as the tympanic membrane. Analyses show that to determine risks caused by exposure to noise sources during the performing of working activities, it is necessary to monitor the phenomena's dynamic of the phenomena through analyses of SPL time histories. Otherwise several situations of risk, which in time may cause irreversible damage to workers health, may go undetected. From the experience gained with the analyses carried out, it is evident that acoustic-characteristics of a work place are poorly represented when the descriptors indicated in the Directive (p_{peak}, $L_{EX, 8h}$, $L_{WE, 8h}$) are used. For instance, the use of the only peak sound pressure p_{peak} leads to non-identification of possible components with significantly high energetic content probably harmful to the human ear. This leads to one of the main lapses in the currently used legislation; no importance is given to the impulsive components. They are only considered in the evaluation of the overall exposure level (either $L_{EX, 8h}$ and $L_{WE, 8h}$). Because they are characterized by high pressure levels, but a short duration (a few milliseconds), they do not increase the continuous equivalent SPL of the signal. Therefore, their contribution, is not considered at all and no protective measurements are taken.

The second weakness of the Directive is the choice of the C-frequency weighted curve for the evaluation of the peak pressure levels. It artificially decreases the energy value of the signal at the low and very high frequencies, where the protection mechanisms of the human ear are less effective. Weighing operations modify the energy content of the studied phenomenon and this, as shown in this chapter, can dangerously underestimate the effective stress to which the ear organ is subjected to in terms of pressure applied on the internal ear.

The analysis procedures here proposed consist of an integration of the existing norm. These implementations could permit technicians to highlight the presence of certain events and phenomena, which are extremely harmful to hearing and the health of the ear (e.g. impulsive, periodic high-energy events, etc.) that cannot be recognized when the Directive 2003/10/EC is used. Furthermore, the new proposed method suggests to no longer use the C-frequency weighted scale, which artificially decreases the potential danger of noise at low and high frequencies. It is clear that if an effective protection has to be achieved, the whole sound spectrum has to be considered. Conducing analyses using an un-weighted signal allows for a more conservative, and likely more protective action to ensure worker health care and protection from damage arising from prolonged exposure to noisy processes.

ACKNOWLEDGMENTS

This work and the measurements presented in this chapter have been made possible thanks to the collaboration and the availability of the administrative staff of the Dipartimento di Ingegneria Meccanica, Energetica e Gestionale

(DIMEG) of the Università della Calabria, Italy and the Laboratorio dei Materiali of the same Department. In addition, the authors would like to thank Ing. Gianfranco Marcelli, Director of A.M.A.CO. S.p.A, for the availability shown and for have made the measurements carried out on their vehicles possible.

REFERENCES

[1] Stansfeld, S. A. and Matheson, M. P., "*Noise pollution: non-auditory effects on health*", *British Medical Bulletin*, Vol. 68, Iss. 1, pp. 243 - 257, 2003.

[2] Ising, H.and Kruppa, B., "*Health effects caused by noise: Evidence in the literature from the past 25 years", Noise & Health*, Vol. 6, Iss. 22, pp. 5 – 13, 2004.

[3] Directive 2003/10/CE of the European Parliament and of the Council of 6 February 2003 – *The minimum health and safety requirements regarding the exposure of workers to the risk arising from physical agents (noise).*

[4] Reda, A. and Sabato, A., "*Innovative risk assessment with impulsive events*", in Proc. Internoise 2006, Honolulu – HI, pp. 1425 - 1432, 2006.

[5] Passchier-Vermeer, W. and Passchier, W. F., "Noise exposure and public health", *Environmental Health Perspective*, Vol. 108, Iss. 1, pp. 123 – 131, 2000.

[6] Job, R. F. S., *"The influence of subjective reactions to noise on health effects of the noise", Environment International*, Vol. 22, Iss. 1, pp. 93 – 104, 1996.

[7] Davis, H., Morgan, C. T., Hawkins Jr., J. E., Galambos, R., Smith, F. W., "Temporary deafness following exposure to loud tones and noise", *Acta Oto-Laryngologica,* Vol. 88, pp. 56 - 67,1950.

[8] Moore, B., "*An introduction to the psychology of hearing*", 6th edition Ed. Emerald Group Publishing Limited, Bingley – UK, 2012.

[9] Glorig, A., "The effects of noise on man", *The journal of the American medical association,* Vol. 196, Iss. 10, pp. 839 – 842, 1966.

[10] Gray, H., "Anatomy of the human body" - *The Organs of the Senses and the Common Integument*, Lea & Febiger, Philadelphia – PA, pp. 1029 – 1057, 1918.

[11] Yost, W. A., "*Fundamentals of hearing: an introduction*", 5th edition, Ed. Academic Press, San Diego – CA, 2007.

[12] Robles, L. and Ruggero, M. A., "Mechanics of the Mammalian Cochlea", *Physiological Reviews*, Vol. 81, 1305 – 1352, 2001.

[13] Anastasi, G., Balboni, G. and Motta, P., *"Trattato di Anatomia Umana"*, Vol. III, Ed. Ermes, Milan - Italy, 2012.

[14] Pickles, J. O., "*An introduction to the physiology of hearing*", 4th edition, Ed. Emerald Group Publishing Limited, Bingley – UK, 2012.

[15] Fritzsch, B., Jahan, I., Pan, N., Kersigo, J., Duncan, J. and Kopecky, B., "Dissecting the molecular basis of organ of Corti development: where are we now?", *Hearing research,* Vol. 276, Iss. 1 – 2, pp. 16 – 26, 2012.

[16] Hammershoi, D. and Moller, H., "Sound transmission to and within the human ear canal", *The Journal of the Acoustical Society of America,* Vol. 100, Iss. 1, pp. 408 - 427, 1996.

[17] Gelfand, S. A., "*Hearing: An introduction to psychological and physiological acoustics*". 5th edition., Ed. Informa Healthcare, New York City - NY, 2009.

[18] Beranek, L. L., "*Noise and vibration control*", Ed. McGraw-Hill, New York City - NY, 1971.

[19] Bergström, B. and Nyström, B., "Development of hearing loss during long-term exposure to occupational noise a 20-year follow-up study", *International Journal of Audiology*, Vol. 15, Iss. 4 , pp. 227-234, 1986.

[20] D. M. 16 Marzo 1998, *Tecniche di rilevamento e di misurazione dell'inquinamento acustico.*

[21] Møller, A. R., "*Hearing: anatomy, physiology and disorder of the auditory system*", Ed. Academic Press, Boston – MA, 2006.

[22] Kinsler, L.E., Frey, A. R., Coppens, A. B., Sanders, J. V., "*Fundamentals of Acoustics*", 4th edition, Ed. Wiley, Hoboken – NJ, 1999.

[23] Howard, M. D. and Angus, J. A. S., "*Acoustics and Psychoacoustics*", 4th edition, Ed. Focal Press, Burlington – MA, 2009.

[24] Fletcher, H. and Munson, W.A., "Loudness, its definition, measurement and calculation", *Journal of the Acoustic Society of America* Vol. 5, pp. 82-108, 1933.

[25] ISO 226:2003 – *Acoustics - Normal equal-loudness-level contours.*

[26] OSHA 1910 - *Occupational Safety and Health Standards.*

[27] ISO 1999:1990 – *Acoustics – Determination of occupational noise exposure and estimation of noise-induced impairment.*

[28] ISO 1999:2013 – *Acoustics - Estimation of noise-induced hearing loss.*

[29] Zhao, Y. M. and Chen, S.S., "Adjustment of dose-response relationship of industrial impulse noise induced high frequency hearing loss with

different exchange rate", *National Medical Journal of China*, vol. 86, Iss. 1, pp. 48 – 51, 2006.

[30] ISO 1996-2:2007 – *Description and measurement of environmental noise. – Part 2: Determination of environmental noise levels.*

[31] D. Lgs 9 Aprile 2008, n. 81, *Testo unico in materia di salute e sicurezza nei luoghi di lavoro.*

[32] [32] Zhao, Y. M. and Chen, S.S., "Relationship between impulse noise and continuous noise inducing hearing loss by dosimeter measurement in working populations", *National Medical Journal of China*, Vol. 39, Iss. 6, pp. 396 – 399, 2005.

[33] *Council Directive 86/188/CEE of 12 May 1986 on the protection of workers from the risks related to exposure to noise at work.*

In: Noise Pollution
Editor: Eugene Roberto Nicchi

ISBN: 978-1-63321-110-0

Chapter 5

OTOACOUSTICS EMISSIONS BY PRODUCT OF DISTORTION IN METALLURGICAL WORKERS EXPOSED TO DIFFERENT DOSES OF OCCUPATIONAL NOISE

***Marlene Escher Boger*[*], *André Luiz Lopes Sampaio*[†], *Monique Antunes de Souza Chelminski Barreto*[≠] *and Carlos Augusto Costa Pires de Oliveira*[δ]**
University of Brasilia, Brazil

ABSTRACT

Introduction: The otoacoustic emissions test is useful in monitoring changes of hearing not yet detectable by tonal audiometry, as well as to monitor the damage in the cochlea caused by exposure to noise. Objective: To investigate the evoked otoacoustic emissions by distortion-product in metallurgical workers exposed to different doses of occupational noise. Method: This was a prospective cohort study

[*] Audiologist, Master of Science Health and Ph.D. in Health Sciences, University of Brasilia, Brazil
[†] Physician, Otorhinolaryngologist, Ph.D. in Health Sciences, University of Brasilia, Brazil
[≠] Audiologist, Master of Science Health and Ph.D. Student in Health Sciences, University of Brasilia
[δ] Physician, Otorhinolaryngologist, Ph.D. in Medicine, University of Brasilia

conducted at normal hearing metallurgical workers. The evaluated subjects were divided in three groups: not exposed to noise (Group I), sporadically exposed to noise (Group II) and often exposed to occupational noise (group III). Results: There were significant changes in otoacoustic emissions in Groups II and III, considering both ears. In the amplitude and signal/noise ratio it was observed that the higher the frequency of the sound, the worse the results in Groups II and III, and the greater the exposure to noise, the lower the average found in otoacoustic emissions. The frequency with the worst records was 6 kHz. The Group II, although sporadically exposed to noise, suffered the effects of exposure and revealed a major risk for hearing loss. Conclusion: Evoked otoacoustic emissions were shown to be a sensitive method to detect changes in cochlear function in both groups.

Keywords: otoacoustic emissions, occupational noise, workers

INTRODUCTION

The test of otoacoustic emissions (OAE) has been highlighted among the set of methods to assess hearing, due to its characteristics of speed, objectivity and, above all, the ability to detect early cochlear changes resulting from exposure to noise, not identified by tonal audiometry [1]. Studies indicate that, when there are changes in outer hair cells (OHC), even when there is no change in tone thresholds, it is possible to observe a reduction of OAE amplitude, demonstrating the sensitivity of this parameter in the early diagnosis and/or in accompanying chemotherapy treatments as well as on issues of worker health [2].

In the scientific literature, there are many issues that relate evoked otoacoustic emissions to workers exposed to ocupacional noise [3-11]. However, most of these studies involve only the population of the manufacturing sector, with the working day of 8 hours daily, dividing the studied group of workers at risk and other unexposed population. This fact justifies the importance of studying a third group, those who are exposed sporadically without the use of hearing protectors in order to test the effectiveness of otoacoustic emissions in the investigation of possible irreversible damage that even short exposures to noise in the workplace may cause. The objective of this study is to investigate the distortion-product evoked otoacoustic emissions in metallurgical workers exposed to different doses of occupational noise.

METHOD

This is a prospective cohort study in which participants are classified as exposed, sporadically exposed and not exposed to a particular factor of interest (noise), formed by normal hearing subjects, and conducted in four 4 (four) industries of metallurgical branch. This population was assessed and selected using a convenience sample, made up only by those companies that agreed to participate. These companies are geographically located at the Federal District – Brazil.

The participants of both sexes , aged 18-35 years who presented hearing thresholds equal to or less than 25 dB(A), normal hearing and no history of previous occupational noise exposure or ototoxic chemicals, did not have diabetes, hypertension, vertigo , nor made any ear surgery, cancer treatment or thyroid disease. It was excluded from the study the users of drug for depression, steroids and anti-inflammatory. Data were obtained through occupational history and information from the human resources of the companies that provided medical and laboratory results of periodic health examinations.

The environmental assessment was performed in two stages . At first, it was investigated the distribution of the intensity of the sound production of one metallurgical industry assessed. For this, it was used the SIP 95 level sound pressure meter (sonometer), properly calibrated and located in central points of the factory. At the second time it was evaluated the daily dose of noise exposure. The concept of noise dose is used to characterize the occupational noise exposure, expressed as a percentage of sound energy, with reference to the noise dose greater than 100 %, featuring exposure limit exceeded.

To calculate the daily noise dose related to working hours, levels of continuous or intermittent noise were measured in decibel (dB) with a tool for sound pressure level deployed at a compensation circuit at "A" and slow response circuit (SLOW). Readings were taken near the ear of the worker. For the found values for intermediate level of noise was considered the maximum allowable daily exposure relative to the immediately higher level [12].

When, during the workday, occurred two or more periods of noise exposure at different levels, their combined effects were considered, so that if the sum of the fractions C1/T1 C2/T2 + + C3/T3 ... + Cn/Tn exceeds an unity, we considered the exposure above the tolerance limit . In the above equation Cn indicates the total time during which the worker is exposed to a specific noise level and Tn indicates the maximum allowable daily exposure to this level, set by Brazilian legislation.

The apparatus used for this analysis was the "Quest" dosimeter, model DL NoisePro Dosimeter/Q40 often ranging from 10-10000 kHz, distortion less than 5 % and accuracy of ± 1% . This analysis occurred in two (2) of the four (4) companies visited.

Were selected for this analysis active workers in the production of the metallurgical industry sectors, exposed to noise throughout the workday and use of hearing protection (Group III).

The technical criteria for the calculation of noise dose used in this study are similar to those of the Ministry of Labour and Employment. The environmental assessment was performed by a technical team responsible, formed by specialized safety engineers.

The audiometric evaluation was preceded by anamnesis, otoscopy and acoustic rest of 14 hours. This review was carried out to check the minimum response level for frequencies from 250 Hz to 8 kHz. The equipment used in this evaluation were: otoscope Welch Allyn, with WA accessories; Interacoustics clinical audiometer model AC30; audiometric booth Redusom acoustic, model RO - 80 Std. All workers were evaluated and those who had normal hearing (thresholds equal or lower than 25 dB(A)) and met the inclusion criteria underwent distortion-product otoacoustic emissions evoked (DPOAE) analysis.

The otoacoustic emissions test was conducted in a sound-treated room, with the aim of specifically evaluating the functionality of outer hair cells. It was established that the right ear of each individual were the first to be tested at this stage. The otoacoustic emissions equipment used in this study, was the MAICO Ero-Scan, which automatically monitored the noise level, the linearity of the stimulus during the test, and the adequate positioning of the probe.

To indicate the time at which each of these aspects has become inadequate for testing, appeared on the screen, respectively, the messages "NOISY" and "NO SEAL". To solve the ear plug was exchanged or repositioned, and the assesment resumed. DPOAE were evaluated through simultaneous presentation of two different pure tones (F1 and F2), expressed by the ratio of 1.22. It was used the parameter of intensity L= 65 and L= 55 dB, cochlear conditions being measured at frequencies of 1.5 kHz, 2 kHz, 3 kHz, 4 kHz ,5 kHz and 6 kHz. The analyzes were made by DPOAE frequency, following the greater than -5 dB amplitude criteria (DP) and the relationship between signal/ noise ratio (SR) greater than 6 dB [13]. The occurrence of DPOAE response at one frequency was considered when it was observed the determined values set in the two above criteria. The criteria of this research is based on recent studies, in subjects with various types of sensorineural hearing loss, which have nominated as the best DPOAE protocol test, and the relationship of L1>L2, or more specifically, L1 = 65 and L2 = 55 dB14 -17.

To conduct the survey three groups of subjects, divided as follows, were evaluated:

Group I: 50 normal-hearing subjects, according to the inclusion criteria, individuals with similar characteristics from the other groups, but not exposed to occupational noise.

Group II: 50 normal-hearing subjects exposed to noise sporadically. This group consisted of active workers in the administrative areas of the metallurgical industries, underwent a short exposure to noise during the workday, moving around the industry without using personal protective equipment headset as the reports presented in the introduction of this study. This population was from the management sector, medical clinic, engineering, warehouse, quality and safety.

Group III: 50 normal-hearing subjects, users of hearing protective equipment, often exposed to noise levels above 85 dB for eight hours daily, for periods ranging between 6 and 18 months. This group remains in production factories (assembly, maintenance and tooling) sectors, acting directly on noisy machines such as forklift, mobile crane, sander, mechanical shears, cutting machine, squeezer and pneumatic hammer. Workers in the coatings industry, known as roller coating were excluded from the study due to concomitant exposure to organic solvents.

The results between Groups I and II, Groups I and III and finally, Groups II and III were compared.

Metallurgical companies invited to participate in the study are large and aware of the importance of using personal protective equipment . Note also that employees who work in the production of industry sectors are rotated among the instruments of labor, i.e. manipulate all machines equally for equal periods and avoiding the repetition of physical movements and continued exposure to a single tool of work that can be more or less noisy compared to others in the same industry. It is observed that with it was ensured that all of this particular group were exposed to the same daily dose of noise during the workday.

The data collected were transported to spreadsheets and descriptive analysis was performed of these: frequency, measure of central tendency (mean) and measure of variance (standard deviation), presented by means of tables. Possible associations between variables were evaluated using the chi-square test. The comparison of the amplitudes and signal/noise ratio (S/N) was performed using an ANOVA mixed design with the factor Group (3 levels) as measured between groups (independent variable) and frequency (6 levels) and laterality factors Ears (2 levels) as repeated measures. The post hoc analyzes were performed using one-way ANOVA or paired t-test second need. The level of statistical significance was set at 5 % ($p < 0.05$).

This work was submitted to the Research Ethics Committee of the Human Sciences College of Health, University of Brasilia, registered 113/10 and approved at the 9th Annual Meeting (05/10/2010).

Results

It were made 608 auditory threshold evaluations (audiometry), preceded by anamnese and otoscopy. Of these, 150 tests complied with the inclusion criteria and were distributed as follows : Group I: 50 individuals not exposed to noise, Group II: 50 subjects exposed to occupational noise sporadically and Group III: 50 subjects exposed to occupational noise with daily use of Personal Protective Equipment (PPE) headset, totaling 300 ears evaluated with the DPOAE test. The study subjects had a mean age of 26.2 years (SD ± 5.7) and 80% were male. The female participants were distributed between Groups I and II. Already in Group III (often exposed) all subjects were male.

Environment Characteristics

By the first environmental analysis (sonometer) it was found that the sound pressure levels of the manufacturing sector in the metallurgical industry showed an equivalent sound pressure level (Leq) of 85.4 dB (A).

In evaluating the noise dose measured in Group III (exposed to noise), it was noted the presence of a substantial risk of occupational hearing loss because the dose exceeded the reference value of 100% (Table 1).

Hearing Characteristics

It was observed that 40% of 150 subjects evaluated had abnormalities in both ears. Evaluating each ear separately, 43.3 % of failures occurred in the left ear and 45.3 % in the right ear, featuring lesion in outer hair cells of the cochlea. There was no statistically significant association between laterality of ear (left or right) and the presence of failure.

Table 1. Dose and time of maximum tolerated exposure in metalurgical workers

	Exposure levels obtained in measurements				
Metalurgical	Time of measurement	Sector	dB(A)	Dose (%)	Time Tolerated
Factory 1	15 min	Production	92,4	279,9	02h40min
Factory 2	15 min	Production	90,1	203,2	03h00min

For all three groups, it was found that, both the left ear and the right ear, the group of workers exposed daily (GIII) had the highest prevalence of abnormalities (66% and 72% respectively).

However, we emphasize that the results obtained in GII (exposed sporadically) revealed high prevalence of abnormalities (50% in the left ear and right ear in 48%) (Table 2).

Table 2. Prevalence of DPOAE changes in groups I, II and III according to the test result and laterality

CHANGES TO THE DPOAE – LEFT EAR								
	GI		GII		GIII		Total	
Outcome	N	%	N	%	N	%	N	%
Pass	43	86,0	25	50,0	17	34,0	85	56,6
Failure	07	14,0	25	50,0	33	66,0	65	43,3
Total	50	100,0	50	100,0	50	100,0	150	100,0
CHANGES TO THE DPOAE – RIGHT EAR								
	GI		GII		GIII		Total	
Outcome	N	%	N	%	N	%	N	%
Pass	42	84,0	26	52,0	14	28,0	82	54,6
Failure	08	16,0	24	48,0	36	72,0	68	45,3
Total	50	100,0	50	100,0	50	100,0	150	100,0

Comparing the variables it was found a statistically significant association between the group of participants and the presence of change in DPOAE test of the left ear ($\chi 2 = 28.887$, df = 2, $p < 0.001$) and right ear ($\chi 2 = 31.851$, df = 2, $p < 0.001$).

Compared to the overall average of the amplitudes, laterality of ears and frequencies tested, it was observed that with increasing frequency, there was a decrease in amplitudes. It was also found that from 3 kHz as the subjects are exposed to occupational noise, the mean decrease, that is, the larger the exposure, smaller is the amplitude found. Thus, groups of exposed workers (GIII) and exposed sporadically (GII) had worse records than the group of subjects not exposed (GI).

Table 3. Overall Average DPOAE amplitudes according to laterality and the frequencies in groups I, II and III

DPOAE AMPLITUDE - LEFT EAR						
	GI		GII		GIII	
Freq.(kHz)	Average	±SD	Average	±SD	Average	±SD
1,5	8,1	5,7	4,7	6,1	2,6	6,0
2	2,9	4,3	4,3	5,5	3,6	6,1
3	2,9	5,3	0,6	6,1	0,6	6,7
4	2,8	6,3	-1,7	6,5	-3,2	6,4
5	2,9	6,0	-3,4	6,9	-3,1	7,2
6	-1,6	5,5	-3,8	8,6	-4,1	8,6
DPOAE AMPLITUDE - RIGHT EAR						
	GI		GII		GIII	
Freq.(kHz)	Average	±SD	Average	±SD	Average	±SD
1,5	9,4	5,9	4,9	5,3	3,2	6,2
2	3,0	4,3	4,7	6,5	4,2	6,7
3	1,3	4,9	1,0	6,5	0,5	7,3
4	2,8	4,3	2,0	6,4	-3,0	6,2
5	4,6	5,9	0,1	7,2	-2,8	6,5
6	0,1	6,1	-4,8	8,4	-3,7	7,8

±SD – standard deviation.

Nevertheless, it is possible to notice that, for the left ear, the Group II had lower amplitude at frequency of 5 kHz (-3.4) when compared to Group III (-3.1) and for the right ear at 6 kHz. This finding was also observed (Group II -4.8 and Group III -3.7) (Table 3). In this case, Groups II and III showed no statistically significant differences.

Mixed design ANOVA factor found no significant effect of group ($F_{2, 146}$=10,513, p<0,001). The procedure for multiple comparisons found that the GI was significantly larger mean amplitudes compared to GII and III (p=0.001 and p < 0.001 respectively) regardless of frequency or ear evaluated. The laterality factor had no statistically significant effect on the mean amplitude ($F_{1, 146}$=0,631, p=0,428).

Table 4. Average amplitudes of changes to DPOAE, according to laterality and the frequencies in groups I, II and III

AMPLITUDES OF CHANGES TO DPOAE - LEFT EAR						
	GI		GII		GIII	
Freq.(kHz)	average	±SD	average	±SD	average	±SD
1,5	6,5	7,1	4,4	6,3	2,6	6,1
2	5,6	3,6	4,1	5,9	3,6	6,1
3	0,5	3,4	0,3	6,2	0,6	6,7
4	0,5	3,8	-2,2	6,6	-3,2	6,3
5	1,2	2,9	-3,8	7,3	-3,1	7,2
6	-6,1	5,8	-7,3	9,0	-8,1	8,8
AMPLITUDES OF CHANGES TO DPOAE - RIGHT EAR						
	GI		GII		GIII	
Freq.(kHz)	average	±SD	average	±SD	average	±SD
1,5	9,9	3,1	4,3	5,4	3,2	6,2
2	5,3	3,9	4,7	6,7	4,1	6,7
3	2,6	4,0	1,0	6,7	0,5	7,3
4	2,3	3,9	-2,3	6,5	-2,9	6,2
5	4,2	4,7	-4,4	7,3	-2,9	6,5
6	-5,3	4,3	-6,1	8,6	-7,9	7,8

±DP – standard deviation.

The frequency factor also showed a significant effect on the mean amplitude ($F_{5, 730}$=96,249, p<0,001) regardless of the group or the assessed ear. All frequencies differ from each other (p<0.001). It was also found a significant interaction between group and frequency factors ($F_{10, 730}$=10,268, p<0,001). The *post hoc* analysis compared the means of each group at each frequency studied. At frequencies 1.5, 4 and 5kHz the average obtained by GI was significantly higher than that obtained by other groups (p<0.015 in both cases). The Ear × Group, Frequency × Ear, and Ear × Group interactions showed no significant effects (F's<1,758, p's>0,069).

Table 5. General average to the DPOAE signal/noise ratio according to laterality and the frequencies in groups I, II and III

DPOAE SIGNAL/NOISE RATIO - LEFT EAR						
	GI		GII		GIII	
Freq.(kHz)	average	±SD	average	±SD	average	±SD
1,5	15,3	7,0	12,5	7,3	11,9	7,1
2	20,6	6,6	13,6	6,8	13,7	6,7
3	21,3	6,2	14,5	7,5	15,8	8,3
4	14,6	7,7	15,4	6,9	14,1	6,5
5	15,5	6,8	13,1	7,8	14,2	8,8
6	9,9	5,4	13,6	8,7	13,6	8,6
DPOAE SIGNAL/NOISE RATIO - RIGHT EAR						
	GI		GII		GIII	
Freq.(kHz)	average	±SD	average	±SD	average	±SD
1,5	15,9	5,8	11,7	5,3	13,1	6,1
2	21,1	4,6	13,6	7,4	14,0	7,8
3	14,4	5,6	15,5	6,9	16,1	7,6
4	20,9	4,4	15,3	6,5	14,2	6,2
5	16,4	6,7	12,8	6,9	13,4	8,4
6	13,5	8,2	12,3	8,3	13,9	7,9

±DP – standard deviation.

When considered only the amplitudes of subjects with altered DPOAE, it is observed that the worst averages were found in the frequency of 6 kHz in the three groups on both the left ear and the right with smaller amplitudes than -5 , pointing especially at this hearing damage frequency. When comparing the groups together, the Group II shows smaller amplitudes than the Group III at the frequency of 5 kHz in both ears, revealing that sporadic exposure to occupational noise interferes DPOAE (Table 4). However, comparisons of means between Groups II and III showed no statistically significant differences between them.

Regarding the overall DPOAE average in the signal/noise ratio, laterality and the evaluated frequencies, it was found that in both ears and all frequencies, the averages are higher than 6 dB, which represents normality criteria (Table 5).

Mixed design ANOVA factor found no significant effect of group ($F_{2, 147}$=5,601, p=0,005). The procedure for multiple comparisons showed that the GI was significantly higher than the average of GII and GIII (p = 0.008 and p = 0.021 respectively) average S/N.

Table 6. Changed DPOAE signal/noise ratio, according to laterality and the frequencies in groups I, II and III

CHANGES TO DPOAE SIGNAL/NOISE RATIO – LEFT EAR						
	GI		GII		GIII	
Freq.(kHz)	average	±SD	average	±SD	average	±SD
1,5	13,0	5,1	12,7	7,5	11,1	7,1
2	16,3	6,6	13,3	7,1	13,4	6,6
3	18,2	5,2	14,3	8,0	14,7	8,1
4	13,1	6,2	12,9	7,1	13,1	6,6
5	14,1	7,9	12,6	8,1	13,9	9,2
6	4,1	7,9	2,8	8,0	1,3	8,8
CHANGES TO DPOAE SIGNAL/NOISE RATIO – RIGHT EAR						
	GI		GII		GIII	
Freq.(kHz)	average	±SD	average	±SD	average	±SD
1,5	13,3	4,8	11,6	5,5	12,2	6,1
2	20,5	3,4	12,6	7,6	13,2	7,6
3	16,7	6,5	14,9	7,0	15,1	7,7
4	19,5	3,4	15,0	6,6	13,3	6,1
5	15,9	6,0	12,4	7,1	12,9	8,2
6	5,0	3,3	2,3	8,6	1,4	7,3

±DP – standard deviation.

The frequency factor also showed a significant effect on the average S/N ($F_{5, 735}$=16,159, p<0,001). The *post hoc* analysis showed that the frequencies 1.5 5 and 6 kHz had significantly higher mean compared to the other frequencies without considering group or ear evaluated (p's<0.045 in all cases).

The interaction of Group × Frequency factors showed significant effect on the average S/N ($F_{10, 735}$=6,903, p<0,001). The *post hoc* analysis showed significant differences between groups at frequencies 1.5, 2, 4 and 6 kHz (p's<0.007 in all cases). At 4 kHz the GI avarege was greater than the GIII (p= 0.006). The laterality factor showed no statistically significant effect on mean S/N ($F_{1, 147}$=0,633, p=0,428).

On the other hand, when considering only the ears with DPOAE changed, it was found that at the frequency of 6kHz all groups had average below 6 dB, and the greater the risk exposure, the results were much worse on both left and the right ear. Comparisons of means between GII and GIII shows no statistically significant differences (Table 6).

Discussion

The possibility to identify early cochlear alterations in normal-hearing employees motivated many scientists to research the auditory effects, caused by occupational noise, through the OAE test. The choice of distortion-product otoacoustic emissions for this research was based on the possibility to evaluate the activity of the cochlea at specific frequencies, providing comprehensive analysis when compared with the otoacoustic emissions that evaluates the cochlea globally.

Recognizing that DPOAEs may represent an important technical resource to prevent hearing loss induced by high sound pressure level (HSPLIHL), we used this procedure to assess hearing in order to research the cochlear conditions of metallurgical workers, because they are professionals working in auditory risk environment.

For this test it is essential that the middle ear is in proper physiological conditions and to emphasize that it is an efficient, fast and objective test for the differential diagnosis and monitoring of HSPLIHL [11,18], considering that audiometry is a subjective method and directly depend on the response of the employee, it may have disadvantages in detecting response [11], and, therefore, subject to influences such as fatigue, pain, discomfort, stress, inattention, misunderstanding the examination, among others. Still, it is considered that audiometry is extremely important as it reveals the type and degree of hearing loss, and such testing should not be disregarded, but complemented by another whose objectivity add data to the result.

The choice of criteria amplitude greater than -5 dB and the signal/noise greater than 6 dB [14] was efficient in detecting changes in outer hair cells in metallurgical workers. Some researchers use more stringent criteria, such as signal/noise ratio greater than 8 dB, and thus emphasize the possibility of even more alarming results.

Regarding the characterization of the sample, it was observed that in Group III all subjects were male. Probably due to the activity carried out by this group, that demands great physical effort in production of metallurgical industries, employing only male workers in these sectors. The objective of this research is not to compare possible differences between the sexes.

Findings in a comparative study between men and women noted that, in the group of men, the reduced DPOAE covered a greater number of frequencies, and concluded that the amplitudes were reduced in both groups, suggesting that the risk is present for both sexes [2].

In auditory characteristics, it is noted that even with normal audiometric thresholds, 40% of workers presented changes in DPOAE in both ears . The high prevalence found in this study is supported by other authors who demonstrate the sensitivity of this method of hearing assessment in detecting early cochlear changes caused by noise exposure, not identified by tonal audiometry [6-11]. When considering the prevalence of DPOAE changes according to laterality, there is no statistically significant association between the right and left ears , suggesting that there is a tendency for lateralization of hearing loss and exposure occurs equally in both ears this hive of activity.

Examining the prevalence of DPOAE changes in groups studied, it was found that the high rates are between Groups II and III (exposed sporadically and often). This finding reveals that belong to the group exposed increases the chance of developing hearing impairment, and the higher is exposure higher is the hearing risk. Moreover, the percentages found suggested that the failure of DPOAE can occur even with normal audiometric thresholds preceding cochlear changes. Although this study presents different characteristics compared to other already made, due to the fact that it included individuals sporadically exposed to noise, it is noted that the scientific literature also shows that groups not exposed obtains highest occurrences of responses in the groups exposed and DPOAEs are more sensitive to detect the effects of noise [2,11].

When analyzing the overall averages of DPOAE amplitudes, it was found that in both ears averages decreased with increasing frequency, as in other studies [4,8]. In the present study, the mean amplitudes of GI were higher than those in GII and GIII analyzed frequencies, in agreement with the above authors, who also found lower values of amplitude noise in the exposed group.

When we consider only the amplitudes of subjects with altered DPOAE, it is observed that hearing damage is found mainly in the frequency of 6kHz, corroborating other findings that show that frequency as one of the first hit as a result of noise exposure occupational [8]. Another finding shows that the Group II presents amplitudes smaller than the Group III in the frequency of 5 kHz in both ears. Despite not showing statistically significant differences between Groups II and III, it is believed that this result points to the Group II should be considered at risk, and preventive measures need to be taken in order to avoid the emergence of HSPLIHL. It is likely that the lack of statistical association is due to the small number of participants in each group.

The amplitudes in the signal/noise ratio of GI were higher than those in GII and GIII. The fact that the unexposed group have larger amplitudes than the exposed group was also observed in other research [8]. The averaged between Groups II and III appear homogeneous, suggesting that the two groups behave similarly in relation to the hearing results. It was also observed that the signal/noise ratio, the hearing damage mainly in the frequency of 6 kHz, is corroborated in the literature scientific [8] noting that this frequency was the most affected by the deleterious effects of noise exposure.

It is important to stress the results for group II (exposed sporadically), in which, in almost all analyzes there was similarity between Groups II and III , and these may present cochlear injury, extra-auditory and hearing effects due to exposure to noise and yet are not considered as a risk group, according to the criteria and standards currently used by the Ministry of Labor and Employment (Brazil).

Another important finding was verified in the evaluation of noise dose measured in Group III, which states the presence of a significant risk of occupational hearing loss. It was observed that the noise levels are above the tolerance limits for the work day of 8 hours per day, that is 85 dB, according to Regulatory Standard 15 . Although the Group III using hearing protection equipment as required by law, the results point to the fact that law enforcement does not always guarantee the prevention of hearing loss as they are still found high prevalence rates of HSPLIHL in Brazilian workers.

Also in relation to environmental assessment, there was discrepancy in the data concerning to the intensity of the sound pressure, and based on the dose assessment, the noise has become more intense when compared to the result obtained by sonometer. This event can be explained by the way the data were collected, since the position of the sound level pressure meter at the time of review stood at a central point in the industry, capturing the general ambient noise. At the evaluation with the dosimeter, the equipment was positioned close to the worker's ear, which in turn remained in direct contact with machinery or circulated next to them, thus capturing higher sound pressure levels, showing that this method is capable reporting the actual noise exposure.

Finally, when comparing the groups together, it was observed that the criteria studied, there were significant differences, and these were evident with GI presenting better, than GII and GIII. Thus, all criteria demonstrated the negative effect of noise exposure on DPOAE GII and GIII.

The analysis and conclusions of this research can contribute as an effective indicator subsidizing the implementation of preventive measures and surveillance actions that can minimize the effects of occupational noise.

Although workers with concomitant exposure to organic solvents were excluded from this study, through literature review, was possible to identify hearing disorders in groups exposed to noise and solvents [19]. However, studies involving the use of EOA to monitor the hearing of workers exposed to noise and chemicals are still scarce, which makes it essential to have a greater number of investigations concerning several simultaneous exhibitions in industrial environments, in order to confirm the effectiveness of otoacoustic emissions and the importance of the exam together with audiometry. It is believed that otoacoustic emissions are an important ally in the early identification of adverse effects caused by the combination of solvents and noise [20].

Conclusion

Evoked distortion-product otoacoustic emission in normal hearing workers, exposed to different doses of occupational noise, proved to be an effective method for early detection of cochlear changes. These changes were evident in the exposed groups compared to the group not exposed to noise.

References

[1] Marques FP, Costa EA. Exposure to occupational noise: otoacoustic emissions test alterations. *Braz. J. Otorhinolaryngol.* 2006; 72(3):362-6.

[2] Frota S, Iório MCM. Distortion product otoacoustic emissions and pure tone audiometry: a study of temporary threshold shifts. *Braz. J. Otorhinolaryngol. 2002; 68(1):15-20.*

[3] *Barros SM da S,* Frota S, Atherino CCT, Osterne F. The efficiency of otoacoustic emissions and pure-tone audiometry in the detection of temporary auditory changes after exposure to high sound pressure levels. *Braz. J. Otorhinolaryngol.* 2007; 73(5): 592-8.

[4] Salazar A, Fajardo L, Vera C, García M, Solís F. Comparison of otoacoustic emissions produced by distortion in individuals exposed and non-exposed to occupational noise. *Cienc. Trab.* 2003; 5(10):24-32.

[5] Biro K, Nnoszek L, Prekopp P, Nagyiványi K, et al. Characteristics and risk factors of cisplatin-induced ototoxicity in testicular cancer patients detected by distorsion product otoacoustic emission. *Oncol.* 2006; 70(3):177-84.

[6] Korres GS, Balatsouras DG, Tzagaroulakis A, Kandiloros D, et al. Distortion product otoacoustic emissions in an industrial setting. *Noise Health.* 2009; 11(43): 103-10.

[7] Seixas NS, Kujiawa SG, Norton S, Sheppard L, et al. Predictors of hearing threshold levels and distortion product otoacoustic emissions among noise exposed young adults. *Occup Environ Med.* 2004; 61:899-907.

[8] Atchariyasathian V, Chayarpham S, Saekhow S. Evaluation of noise-induced hearing loss with audiometer and distortion product otoacoustic emissions. *J. Med. Assoc. Thai.* 2008; 91(7):1066-71.

[9] Guida HL, Morini RG, Cardoso ACV. Audiologic and otoacoustic emission evaluation in individuals exposed to noise and plaguecides. *Intl Arch. Otorhinolaryngol.* 2009; 13(3):264-9.

[10] Oliveira PF, Raposo OFF, Santos ACA, Santos LA. Otoacoustic emissions as an instrument of epidemiological surveillance in the health of the workers. *Intl. Arch. Otorhinolaryngol.* 2011; 15 (4):444-9.

[11] Baradarnfar MH, Karamifar K, Mehrparvar AH, Mollasadeghi A, et al. Amplitude changes in otoacoustic emissions after exposure to industrial noise. *Noise Health.* 2012; 14: 28-31.

[12] ISO 1999 – Acoustics – Determination of occupational noise exposure and estimation of noise-induced hearing impairment. 1990. *Ero-Scan. Operation Manual MAICO.* 2005; ver. 11.

[13] Gorga MP, Neely ST, Dorn PA. Distortion product otoacoustic emission test performance for a priori criteria and for multifrequency audiometric standards. *Ear. Hear.* 1999;20(4):345-62.

[14] Mills DM. Interpretation of standard distortion product otoacoustic emission measurements in light of the complete parametric response. *J. Acoust. Soc. Am.* 2002;112(4):1545-60.

[15] Avan P, Bonfils P. Distortion-product otoacoustic emission spectra and high-resolution audiometry in noise-induced hearing loss. *Hear Res.* 2005; 209(2):68-75.

[16] Dreisbach LE, Siegel JH. Level dependence of distortionproduct otoacoustic emissions measured at high frequencies in humans. *J. Acoust. Soc Am.* 2005;117(5):2980-8.

[17] Vinck BM, Van Cauwenberge PB, Leroy L, Corthals P. Sensitivity of transient evoked and distortion product otoacoustic emissions to the direct effects of noise on the human cochlea. *Audiol.* 1999;38:44-52.

[18] Botelho CT, Paz APML, Gonçalves AM, Frota S. Comparative study of audiometrics tests on metallurgical workers exposed to noise only as well as noise associated to the handling of chemical products. *Braz. J. Otorhinolaryngol.* 2009; 75(1):51-7.

[19] Martins VV. Otoacoustic emissions in metalworkers with simultaneous exposure to noise and solvents. [dissertation]. Pontifícia Universidade Católica de São Paulo: 2012.

In: Noise Pollution
Editor: Eugene Roberto Nicchi
ISBN: 978-1-63321-110-0

Chapter 6

EARMUFFS NOISE LEAKAGE

***Paulo Henrique Trombetta Zannin*[1,*]**
***and Samir N. Y. Gerges*[2]**

[1]Laboratory of Environmental and Industrial Acoustics and Acoustic Comfort, Department of Mechanical Engineering, Federal University of Paraná, Brazil
[2]Acoustics and Vibration Laboratory, Department of Mechanical Engineering, Federal University of Santa Catarina, Brazil

ABSTRACT

The use of personal ear protection is the only immediate solution for high noise level until a noise control action is carried out.

In this chapter the noise leak pathways through the protector to the ear were investigated.

Measurements with artificial head were performed to evaluate the influence of various parameters on the sound attenuation provided by hearing protectors. The parameters evaluated were: 1) Effect of circumaural cushion, 2) Effect of cup internal lining, 3) Effect of cup weight, 4) Effect of headband force, and the 5) Effect of cup vibration.

The results showed that: 1) Coherence Function between cup vibration and sound pressure inside the cup has shown that higher sound energy is transmitted, at low frequencies, due to cup vibrations. Vibration measured axially at the cup outer surface shows mechanical resonances

* Corresponding Author: Paulo Henrique Trombetta Zannin. e-mail: paulo.zannin@gmail.com.

started around 150 Hz, 2) Internal foam lining of the cup is an important component of the earmuffs to avoid acoustic resonances of the cup cavity, 3) The attenuation measured with a complete earmuff and the same without the cushion has shown that the acoustic energy leakage occurs at low frequency bands. This leakage is due to the low mass density of the cushion materials - foam in this case -, 4) As the headband force increases to a maximal deformation of the cushion, attenuation increases, especially at low frequencies. The increase in the headband force is however a limiting factor to the use of the earmuff, as it can get too uncomfortable, 5) Higher attenuation is obtained at low frequencies as the cup weight increases. This is easily explained by the acoustic transmission mass law. Note that as the protector weight increase, other factors, such as comfort, decreases.

The use of hearing protectors is very important to protect workers against high sound pressure levels in the workplace. On the other hand, also very important, is the quality and maintenance of these protection devices. In the case of earmuffs, it was demonstrated that the removal of the internal foam lining generates acoustic resonances of the cup cavity, which substantially reduces the attenuation provided by the protector at high frequencies.

1. Introduction

The use of personal ear protection is the only immediate solution for high noise level until a noise control action is carried out.

The protection to hearing offered by earmuffs depends upon its design, materials used and several physiological and anatomical characteristics of the wearer (Shaw, 1979; Berger, 1980).

The noise attenuation offered by an earmuff is limited by the bone acoustic conduction and leakage of different pathways such as: 1) Transmission through protector material, 2) Leaks around the protector, 3) The effect of protector cup vibration and cavity resonances (Berger, 1980).

In this chapter, results are presented for the noise leakage due to each of the above pathways, excluding the bone conduction since artificial head were used.

Results provided here are relevant for earmuff projects, with the goal of minimizing noise leak and obtaining acceptable sound attenuation.

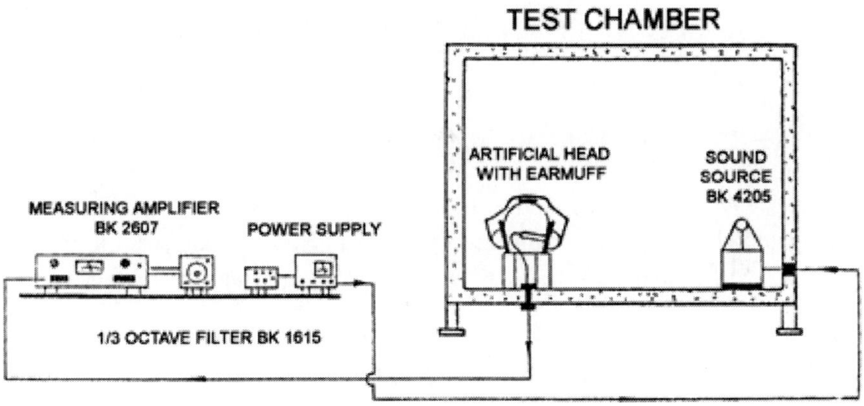

Figure 1. Equipment used in the measurements.

2. Materials and Methods

Measurements were carried out of earmuff attenuation using an artificial head according to ANSI S3.19 (ANSI, 1974), and a diffuse sound field according to ISO 4869-1 (1981) in a test chamber with a volume of approximately 2 m^3 (Figure 1). The measurement of sound attenuation using an artificial head is given by the difference between two sound pressure levels - with and without the earmuff - (ANSI, 1974).

The cast iron artificial head were qualified as well as the reverberation sound field (Zannin, 1990; Zannin and Gerges, 2006).

Three different earmuff models were tested and the results obtained, show a great similarities from the point of view of leakages, thus typical results are presented.

Figure 1 shows the equipment used in the measurements: 1) Test chamber, 2) Sound source - Brüel and Kjaer 4205 (inside the test chamber), 3) Measuring amplifier - Brüel and Kjaer 2607, 4) 1/3 Octave filter - Brüel and Kjaer 1615, 5) Artificial head with earmuff (inside the test chamber), 6) Power supply, 7) ½" Measurement microphone Brüel and Kjaer 4165 (inside the artificial head) (Zannin, 1990).

3. Attenuation of Earmuff Elements

The noise attenuation of a hearing protector depends on the project and materials used in the various elements, such as: cup, muff, foam lining and band force.

The leakage effects of the following pathways and elements, was investigated: 1) Effect of circumaural cushion, 2) Effect of the foam lining, 3) Effect of the headband force, 4) Effect of cup weight, and 5) Effect of the vibration of the cup.

3.1. Effect of Circumaural Cushion

The attenuation was measured using a complete earmuff and using the same earmuff without the cushion, as in Figure 2 (Zannin and Gerges, 2006) and 3, for two different models of protectors. In this case the cup was sealed directly over the artificial head surface.

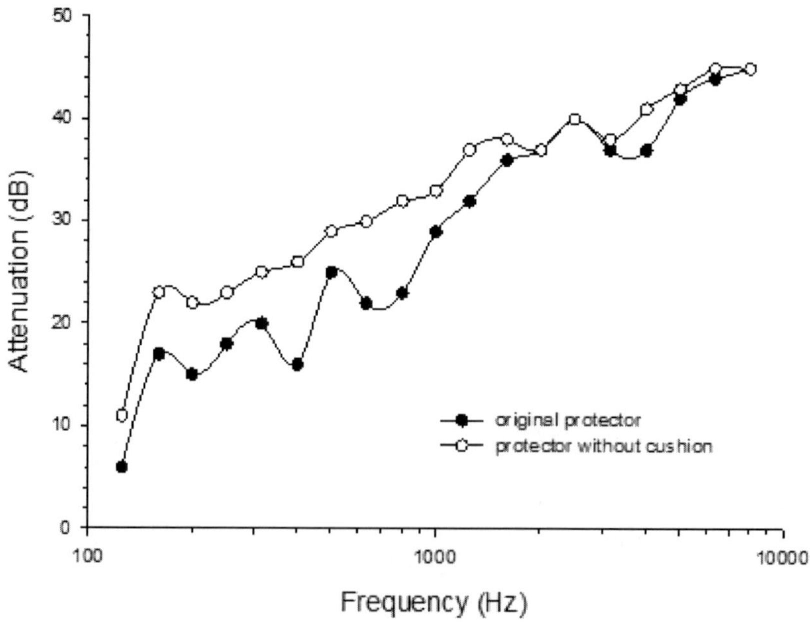

Figure 2. Effect of the cushion on the attenuation of the earmuff.

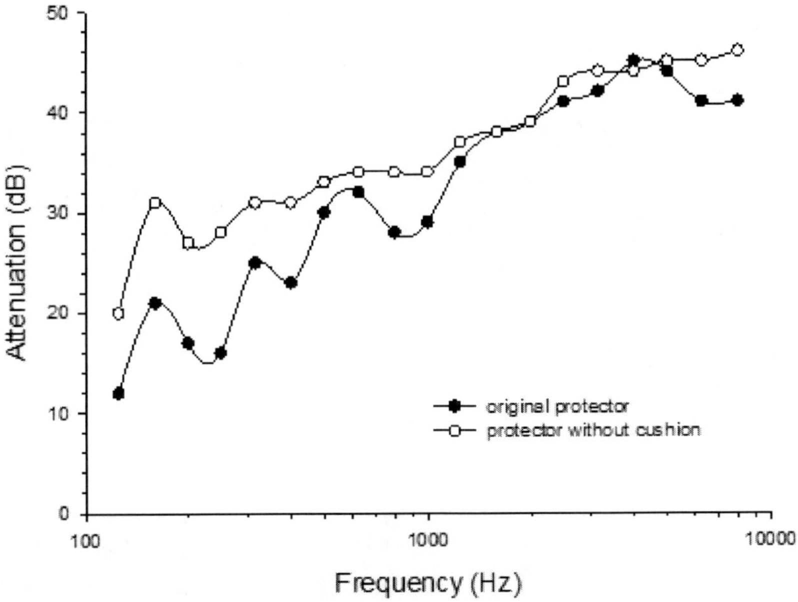

Figure 3. Effect of the cushion on the attenuation of the earmuff.

The results show that the acoustic energy leakage, occur at low and middle frequencies bands. This leakage is due to the low mass density of the cushion materials, in this case - foam - (Shaw, 1979).

3.2 Effect of the Foam Lining

Figure 3 and 4 shows the attenuation measured of a complete earmuff and the same earmuff after removal the internal foam lining of the cup, for two different models of protectors.

The removal of the foam lining of the cup had a clear negative effect on noise attenuation in the frequency of 2500 Hz, and in the frequency range between 4000 and 8000 Hz, as shown in Figure 3.

For the other earmuff, shown in Figure 4, removing the foam lining has negative effects on the attenuation in the frequency range between 3150 and 8000 Hz. At this frequency range, the dimensions of the cup cavity are comparable with a multiple of acoustic half-wavelength hence acoustic resonances of the internal cup cavity occur.

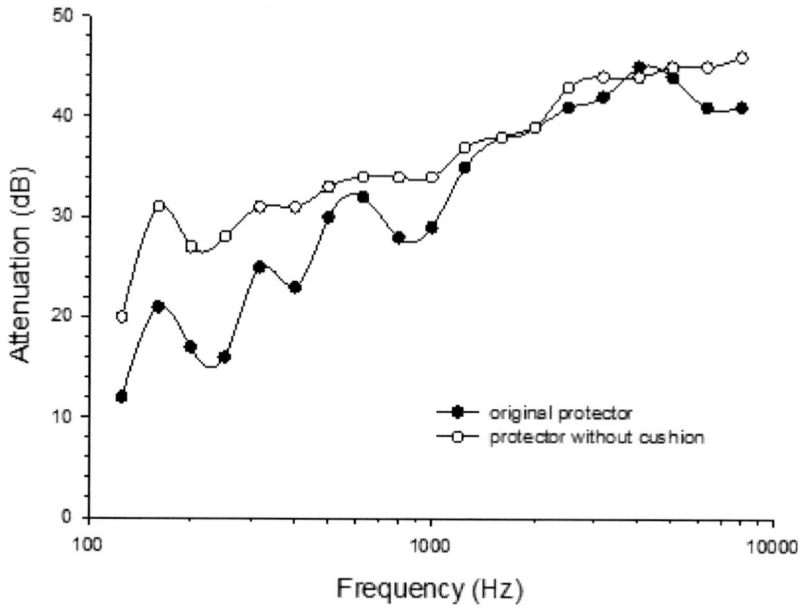

Figure 3. Effect of cup lining on the attenuation of the earmuff.

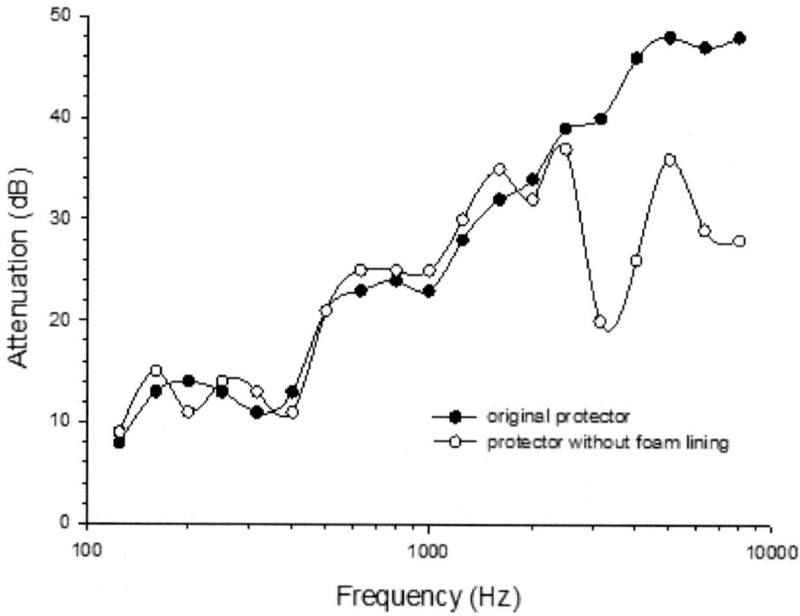

Figure 4. Effect of cup lining on the attenuation of the earmuff.

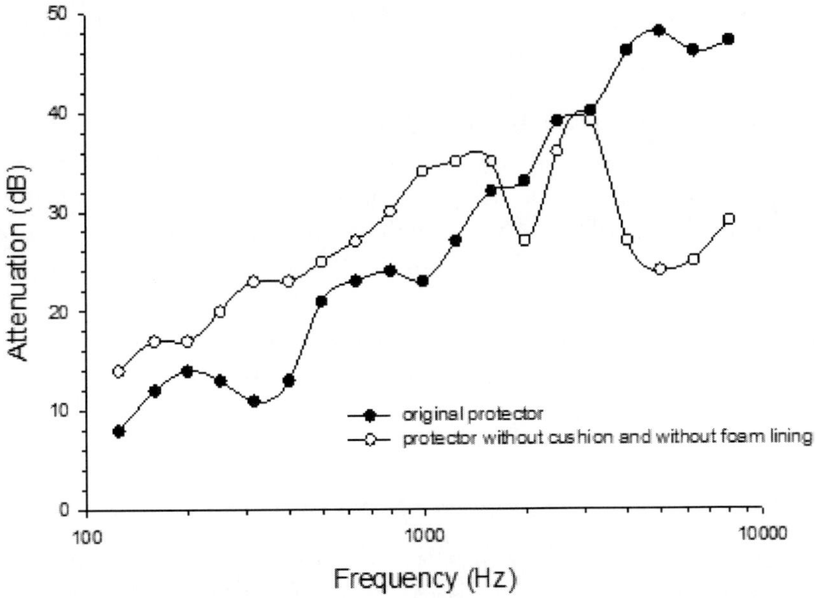

Figure 5. Effect on the attenuation of earmuffs when internal foam lining and cushion are simultaneously removed.

According to Shaw (1979): *"...the modes of the air cavity must be properly damped if the cup is not to behave as a reverberation chamber above 2 kHz"*.

Figure 5 shows the influence on the attenuation of the earmuff, while at the same time, the cushion and foam lining are removed. It is clearly shown that the presence of the cushion exerts a greater negative influence on the attenuation at low and middle frequencies and that the removal of the foam lining exerts a greater negative influence on the attenuation at high frequencies.

3.3. Effect of the Headband Force

Figure 6 shows the sound pressure level reduction obtained for a earmuff with and without an extra headband force - *a force such that the cushion is completely compressed* -.

According to Shaw (1979): *"...only a marginal increase in attenuation can be expected for a large increase in contact force"*.

In contrast to Shaw (1979), Figure 6 shows that as the headband force increases to a maximal deformation of the cushion, a clear increase in the reduction of sound pressure levels is achieved at low frequencies, especially in the frequency range between 100 and 300 Hz.

The increase in attenuation at low frequencies, obtained due to this extra headband force, is caused by the better contact between the earmuff and the lateral surfaces of the artificial head, and by the increase in the mass density of the compressed cushion (Gerges and Zannin, 1992).

The increase in the headband force is however a limiting factor to the use of the earmuff, as it can get too uncomfortable (Berger, 1980).

According to Berger (1980), in real workplace, users will try to diminish the headband force, making the earmuff more comfortable, forcing the headband outwards.

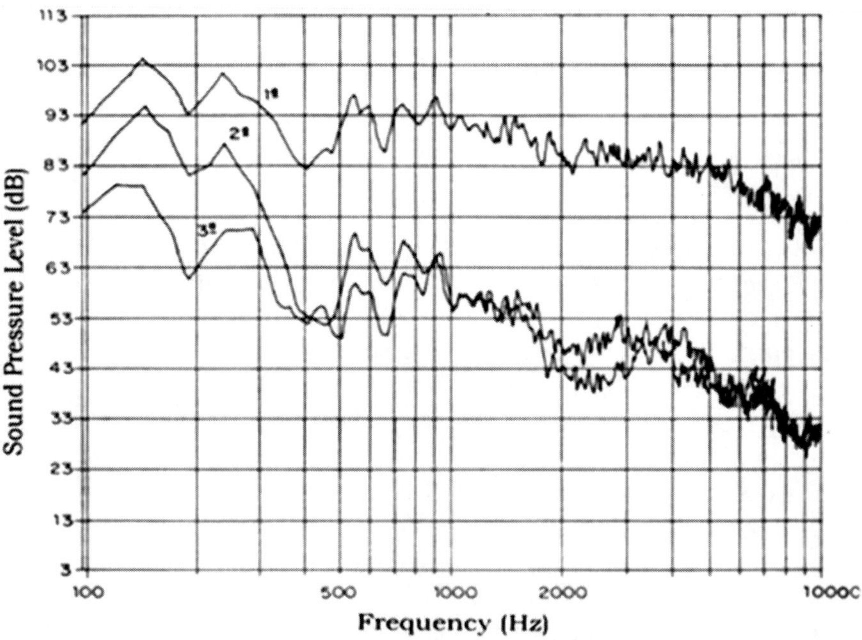

Figure 6. Reduction in sound pressure levels due to the increased contact force: 1) First curve: sound pressure level measured without the earmuff; 2) Second curve: sound pressure level measured using the earmuff; 3) Third curve: sound pressure level measured with a force such that the cushion is completely compressed.

3.4. Effect of Cup Weight

Figure 7 show the reduction in sound pressure levels due to the increased cup weight: 1) Curve 1: Cup weight 59 grams (total protector weight 192 grams), 2) Curve 2: Cup weight 177 grams (total protector weight 428 grams), and 3) Curve 3: Cup weight 295 grams (total protector weight 664 grams).

Figure 7 - curve 2 - shows the increase in cup weight by a factor of three, and - curve 3 - shows the increase in cup weight by a factor of five.

Higher attenuation is obtained at low frequencies as the cup weight increases especially in the frequency range between 100 and 300 Hz. This can be easily explained by the acoustic transmission mass law.

Figure 7 shows too, that further increases in cup weight do not result in useful gain in attenuation, except for low frequencies. Not to be forgotten, however, as the earmuff weight increases, comfort decreases. Therefore, a compromise must be found between the weight and the sound attenuation of hearing protectors.

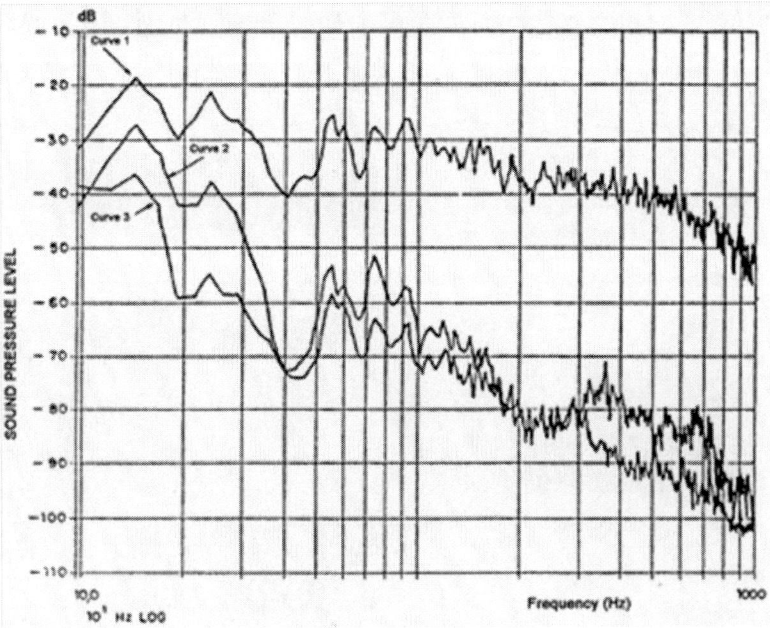

Figure 7. Reduction in sound pressure levels due to the increased cup weight: 1) Curve 1: Cup weight 59 grams (total protector weight 192 grams), 2) Curve 2: Cup weight 177 grams (total protector weight 428 grams), and 3) Curve 3: Cup weight 295 grams (total protector weight 664 grams).

3.5. Effect of the Vibration of the Cup

The vibration of the outer surface of the cup, were measured together with the sound pressure level inside the cup. Figure 8 shows the artificial head with the microphone Brüel and Kjaer 4165 used in the measurements, and the axial and lateral positions of the accelerometer Brüel and Kjaer 4375.

The coherence function between cup vibration and sound pressure is shown in Figure 9, for axial and lateral vibrations (Zannin, 1990; Zannin and Gerges, 2006). These results show that higher sound energy is transmitted, at low frequencies, due to cup vibrations.

The cup and the cushion act as a single degree of freedom mechanical system, in which the mass is given by the cup weight and the spring action is given by the cushion and arc rigidity (Shaw, 1979).

Vibration measured axially at the cup outer surface shows mechanical resonances started around 150 Hz (see Figure 10). At the mechanical resonances the noise transmission loss is very low and a large sound energy is passed to the cup cavity. Lowering these resonance frequencies, the attenuation at low frequencies is increased. This can be achieved by increasing the headband force and/or increasing cup weight (see Figures 6 and 7).

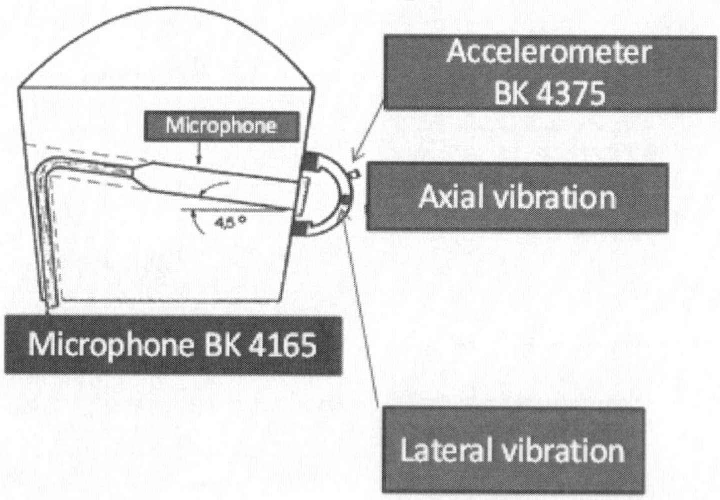

Figure 8. Measurement of the vibration of the outer surface of the cup and the sound pressure inside the cup.

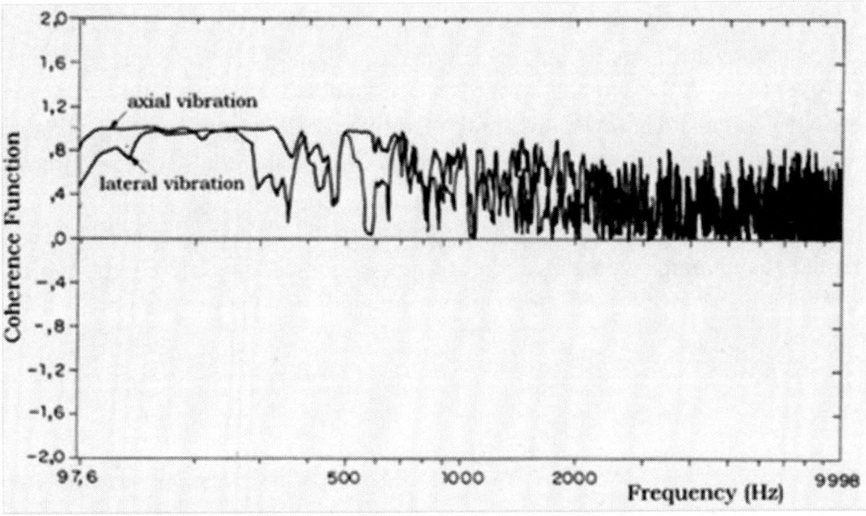

Figure 9. Coherence Function between the vibration of the outer surface of the cup and the sound pressure inside the cup.

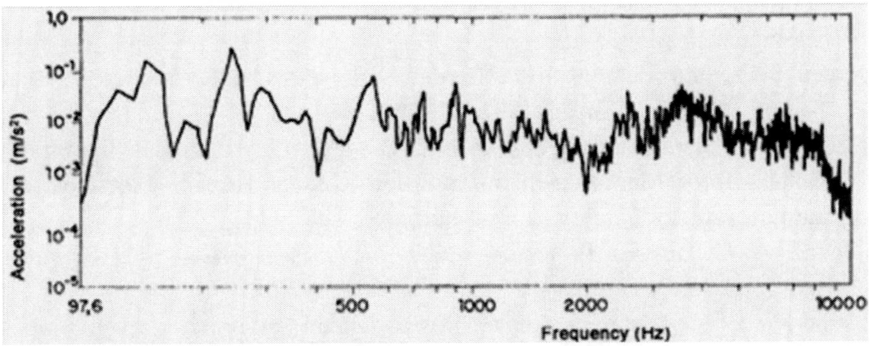

Figure 10. Axial vibration acceleration of cup outer surface.

CONCLUSION

The use of hearing protectors is very important to protect workers against high sound pressure levels in the workplace. On the other hand, also very important, is the quality and maintenance of these protection devices.

The results allow us to conclude that the sound attenuation conferred by earmuffs at low frequencies depend mainly upon: 1) Contact force,

2) Vibration of the axial surface of the cup, 3) Presence of the cushion, and 4) Cup weight. The increase in the contact force and/or cup weight, provide improvements in the attenuation at low frequencies. However, increasing in headband force and cup weight, are limited by protector comfort. Regarding to the cushion, efforts should be made towards finding materials for the cushion which result in less noise leak.

In the case of earmuffs, it was demonstrated that the removal of the internal foam lining, generates acoustic resonances of the cup cavity, which substantially reduces the attenuation provided by the protector at high frequencies.

REFERENCES

American National Standard ANSI S3.19: Method for the Measurement of Real-Ear Protection of Hearing Protectors and Physical Attenuation of Earmuffs, 1974.

Berger EH. E•A•RLog 5, 1980. E•A•RLog is a registered trademark of Aearo Company. Copyright 1996. First printing 1980 (www.e-a-r.com/pdf/hearingcons/earlog5.pdf).

Gerges SNY and Zannin PHT. Earmuffs noise leakage. Inter-Noise 92, Toronto, Ontario, Canada, 1992 July 20-22.

International Standards Organization ISO 4869-1, Acoustics Simplified Method for a Measurement of Insertion Loss of Hearing Protectors of Earmuff type for Quality Control Purposes, 1981.

Shaw E. A. G. Hearing Protector Attenuation: A perspective view. *Applied Acoustics*, 1979, 12, 139-157.

Zannin PHT. Estudo sobre os efeitos dos parâmetros do protetor auditivo tipo concha na atenaução de ruído. Dissertação de Mestrado, Universidade Federal de Santa Catarina, Brasil, 1990. (Master's Dissertation, in Portuguese)

Zannin PHT and Gerges SNY. Effects of cup, cushion, headband force, and foam lining on the attenuation of an earmuff. *International Journal of Industrial Ergonomics*, 2006, 36, 165-170.

In: Noise Pollution
Editor: Eugene Roberto Nicchi
ISBN: 978-1-63321-110-0

Chapter 7

UNDERWATER NOISE POLLUTION: SONIC THREAT OF OFFSHORE WINDFARMS TO MARINE WILDLIFE

Ryunosuke Kikuchi
Faculty of Science and Technology, Ryukoku University, Otsu, Japan
CERNAS/ESAC – Instituto Politécnico de Coimbra, Coimbra, Portugal

ABSTRACT

Today, offshore wind power is a small but growing industry because wind yields offshore are greater than yields for onshore installation. Average wind speeds over water are typically 20% higher than at nearby onshore locations. The most significant concerns relating to wind farms appear to center on noise production. However, (i) only a few studies deal with the impact from offshore wind farms on fish, (ii) knowledge of noise effects on marine life is limited, and (iii) there are significant data gaps. Thus, data on the impact of underwater noise from offshore wind farms on marine life have not yet been compiled in a systematic manner. It therefore seems worthwhile to draw a general picture concerning the effect of underwater noise from wind power turbines on marine life. The following general picture is drawn in this chapter: the short-term potential impact during pre-construction; the short-term intensive impact during construction; and the physiological and/or masking effects that may occur over a long period while the wind farm is in operation. It can be concluded that offshore wind farms should be listed in noise databases in order to promote rational environment management.

1. INTRODUCTION

Although we may all be aware of climate change and air pollution, noise pollution might not get the attention it deserves. Busy roads and planes flying by can be annoying to us, but their impact on wildlife ecosystems should not be overlooked (Harrison, 2014).

It has become increasingly apparent that anthropogenic (human-generated) sound has the potential to affect the health and wellbeing of animals as well as humans (review in Popper and Hastings (2009)). There is also an increasing awareness of the presence of anthropogenic sounds in the aquatic environment and concern has arisen that these sounds could affect aquatic mammals, diving birds, fishes, amphibians, reptiles and perhaps even invertebrates (National Research Council (NRC), 1994, 2000, 2003, 2005).

Despite the concerns raised by the increased presence of anthropogenic sound in the aquatic environment, very little is known about the effects of exposure to such sounds on marine ecosystems (Southall et al., 2007; Hastings, 2008), and far less is known about the effects on fishes (reviews in NRC, 1994, 2000, 2003): one must always be cautious when extrapolating outside the bounds of empirical data, and because data available for the effects of sound on fishes are so few, extra caution is advised during any attempt to extrapolate between fish species, even for identical stimuli. Moreover, one must also be cautious with any attempt to extrapolate results between stimuli because the characteristics of the sources (e.g. air guns, sonars, ships, pile driving) differ significantly from one another (review in Popper and Hastings (2009)). The U.S. Navy's Low Frequency Active Sonar, used to detect submarines, could affect marine life over an area of about 3.9 million Km^2 (Johnson, 2003). This figure is based on levels shown to produce avoidance in fish and whales. Seismic air gun noise from oil and gas exploration measured 3,000 km away was the loudest part of the background noise heard underwater (Nieukirk et al., 2004). In some areas, underwater background noise levels have doubled every decade for the last several decades, most likely due to commercial shipping (McDonald et al., 2006). While any considerable novel addition to an ecosystem (like noise) can be expected to have an impact, one could argue that, because the oceans are naturally noisy environments (wave noise, rainfall, lightning strikes on water, etc.), marine animals would be adapted to noise (Weilgart, 2008). However, such adaptations would only be expected to be for natural noise sources commonly encountered over evolutionary time (Weilgart, 2008). The same would probably not apply to anthropogenic (human-made) underwater noise, which is a relatively recent addition to the marine environment.

Offshore wind power refers to the construction of wind farms in bodies of water to generate electricity from wind. This offshore wind power is a small but growing industry because wind yields offshore are greater than yields for onshore installation. Average wind speeds over water are typically 20% higher than at nearby onshore locations, so an offshore turbine can expect to capture 50% more wind energy than a similar onshore turbine because of the cubic relationship between velocity and power (Reeves, 2003). In light of the efficiency mentioned above, offshore wind power in Europe is expected to increase from 1.1 GW in 2007 to 50 GW in 2020 (Edge and Blanchard, 2007). The most significant concerns relating to wind farms appear to center on noise production (Simmond and Dolman, 2008): however, (i) only a few studies deal with the impact from offshore wind farms on fish, (ii) knowledge of noise effects on marine mammals is limited, and (iii) there are significant data gaps. Thus, data on the impact of underwater noise from offshore wind farms on marine life have not yet been compiled in a systematic manner.

It seems worthwhile to draw a general picture concerning the effect of underwater noise from wind power turbines on marine life. The present chapter aims to discuss the effects of underwater noise from offshore wind power on marine life as the first step of a biological risk assessment. A marine ecosystem is defined as the sum total of marine organisms living in a particular sea area, the interactions between those organisms, and the physical environment in which they interact (Division for Ocean Affairs and the Law of the Sea, 2003). The food chain is therefore an important component of the ecosystem; changes in each component may further destabilize the marine ecosystem. Phytoplankton and zooplankton are generally ranked at lower trophic levels of the marine food chain, fish at intermediate levels, and mammals at higher levels (cf. Pauly, 2003). In the main, this chapter discusses noise effects on fish, taking into account the food chain as a concept.

2. Basic Information

It is reported that most marine animals, particularly marine mammals and fish, are very sensitive to sound, using sound for almost all important aspects of their life including reproduction, feeding, avoiding hazards like predators, and navigation (Tyack and Miller, 2002). Sound can travel long distances underwater, sometimes hundreds or even thousands of kilometers, whereas vision is useful only for tens of meters underwater (review in Weilgart, 2008). Therefore, it is not surprising that much marine life uses sound as its principal sense.

2.1. Sensory System Used by Fish to Hear

Marine species (mammals, fish, and invertebrates in particular) have difficulty using their eyes in deep waters due to low visibility (or less light). Instead, they rely on their hearing to perceive and navigate their environment; hearing ability is thus essential for many marine species (Popper et al., 2003). Fish use two independent but related sensory systems to hear: the inner ear and the lateral line (cf. Figure 1). Together, the inner ear and the lateral line make up the octavo-lateralis system, which provides fish with balance, hearing, and the ability to feel vibrations from a distance (Popper et al., 2003).

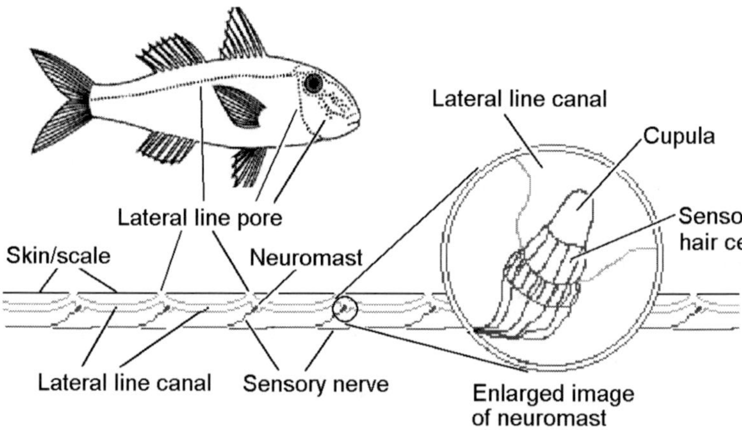

Figure 1. Lateral line (sensory) system used by fish to hear. Sound-induced motion is detected by neuromasts connected with the brain. All fish have some neuromasts. Small canals in the lateral line system are open to the environment via a series of pores in the lateral line. Neuromast organs are located inside the canal. The pores link the outside environment to the fluid inside the lateral line canals where changes in the flow field around the fish are detected. These changes cause the ciliary bundles of hair cells, embedded in the cupula (gel-like cover) on the neuromasts, to bend, alerting the fish to nearby movement (redrawn from Office of Marine Programs (2007)).

2.2. Fish Response to Underwater Sound

Studies have found that fish are affected in a variety of ways depending on the sounds to which they are exposed (review in Office of Marine Programs (2007)). There are currently three basic methods for representing the effects of noise on hearing (cf. Hastings and Popper, 2005). First, the baseline threshold generally represents the lowest signal level an animal will detect in some statistically predetermined percent of presentations of a signal (e.g., 50% of the time). Second, the threshold for temporary changes in sensitivity (temporary threshold shift, abbreviated as TTS) can be measured and treated as a conservative measure of the threshold for injury. Third, permanent threshold shifts (PTS) result in irreversible damage to sensory hair cells (cf. Figure 1). There is an approximately linear relationship between TTS and sound levels above the baseline threshold (Smith et al., 2004a). Compared with the slopes of the lines depicting the relationship for birds and mammals (Figure 2), the slope showing the relationship for fish reflects a narrower dynamic range. If PTS is predicted as the greatest value of TTS, the PTS values for fish are substantially lower than those found for birds and mammals; however, the TTS–PTS relationship for fish has not been studied systematically (review in Hubbs-Sea World Research Institute (2007)). The following typical responses to sound are outlined: hearing loss (physical damage), behavioral changes, and physiological stress.

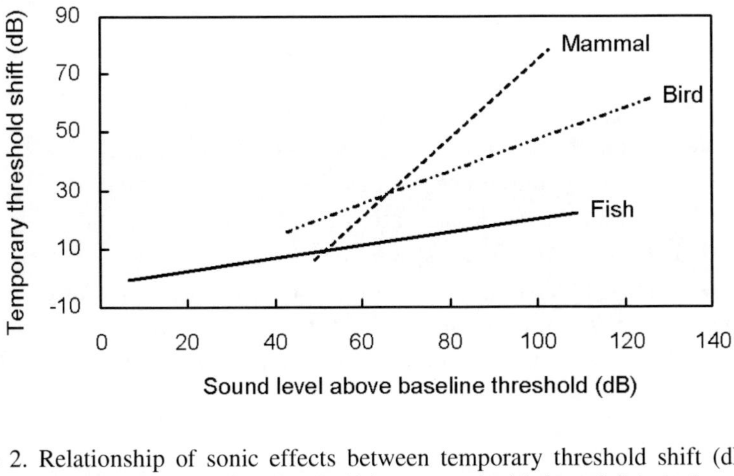

Figure 2. Relationship of sonic effects between temporary threshold shift (dB) and sound level (dB) above baseline threshold in mammals, birds and fish (redrawn from Smith et al. (2004b)).

2.3. Physiological Stress on Fish

Physiological stress responses in fish to anthropogenic sounds have not been documented in the wild (review in Hubbs-Sea World Research Institute (2007) and Office of Marine Programs (2007)). In a laboratory study (Smith et al., 2004b), goldfish were exposed to continuous sounds, and their stress response was quantified by measuring bloodstream levels of glucose and cortisol (a type of hormone produced by the adrenal glands). A short-term spike in cortisol secretion was found during exposure to noise at 160-170 underwater dB. This study is likely the first attempt at investigating the effects of anthropogenic sound on the physiological stress response of a fish species using a stress indicator (cortisol), but additional studies (e.g. noise masking) need to be conducted on other fish species to further understand this effect.

Masking occurs when a loud sound drowns out a soft sound and/or when a noise occurs at the same frequency as a sound signal. It is not known if animals listen to all the sounds they can hear or if they select the sounds to which they want to listen. Masking is influenced by the noise duration. Fish produce broadband transient sounds that have been variously described as clicks, claps, and pops, as well as series of those sounds, which have been described as hums, drumming, grunts, and buzzes. To the extent that fish source levels have been measured, they are usually low to moderate, 70-135 underwater dB (Wahlberg and Westerberg, 2005).

2.4. Behavioral Changes

Several studies have demonstrated the effect of anthropogenic sounds on fish behavior. In two field studies (Engas et al., 1996; Slotte et al., 2004), catch rates of commercially important species (cod, haddock, blue whiting, and Norwegian herring) declined in areas where seismic exploration was conducted with air guns. These studies also show that catch rates of these species increased 30-50 km away from the sound source, suggesting that fish avoid the areas where the air guns are operating. Furthermore, a study of caged pink snapper suggests that fish flee when they hear an approaching air gun array if they are able to swim away (McCauley et al., 2003).

2.5. Noise Injury

A noisemaker test in Jervoise Bay (Western Australia) has shown that loud noise significantly damages fish ears (McCauley et al., 2003). In this test, pink snapper fish were placed in a cage at varying distances from an air gun and exposed to differing levels and repetitions of sound from the gun. When the fish were examined, holes were found in the portions of their ears used for hearing in the region where researchers expected to find hair cells (sensory receptors, see Figure 1). The hair cells had either been ripped away, or there was evidence that the cells were dying. Unlike human hair cells, the hair cells of fish are normally able to regenerate, but there was no evidence of repair/replacement of the damaged hair cells up to 2 months post-exposure. Although fish can freely swim away from a sound source, the test performed on the pink snapper reported that the fish exposed to air gun signals displayed disoriented behavior.

3. Data of Noise Caused by Offshare Wind Farm

The lifecycle of a wind farm (30 years) can be split into four phases (Nedwell and Howell, 2004): (i) the pre-construction phase (lasting 1-5 years), (ii) the construction phase (1 year), (iii) the operational phase (20-25 years), and (iv) the decommissioning phase (1 year). In consideration of this lifecycle, this section reviews available information on underwater noise related to offshore wind farms in the range from pre-construction to operation.

3.1. Ambient Noise in the Ocean

Ambient noise is first briefly reviewed and then followed by a description of windfarm-related noise.

In the frequency range of 500–100,000 Hz, ambient ocean noise is mostly due to spray and bubbles caused by breaking waves. At frequencies greater than about 100,000 Hz, the noise is generated by the random motion of water molecules (so-called thermal noise) (cf. Office of Marine Programs, 2007). Above 100 Hz, wind and waves create sound; this level ranges from 20 to 55 underwater dB (cf. Wenz, 1962). Distant anthropogenic noise begins to dominate between 20 and 80 Hz. Distant shipping traffic is the greatest contributor to man-made ambient noise, with received levels up to 55 underwater dB (Nedwell and Howell, 2004). As ambient noise can vary in magnitude and frequency in both space and time, it may mask wind farm-related noise at some locations and times; however, the masking efficacy must be well below the source under different conditions (Nedwell and Howell, 2004).

3.2. Pre-Construction/Construction Noise

Before construction begins on a wind farm, activities include a geophysical/geotechnical survey of the increase in vessel traffic. Geophysical surveys are typically conducted with boomers and sparkers. A boomer consists of two plates using a high-voltage impulse; the resultant magnetic field vibrates and radiates acoustic energy into the surrounding water. A sparker forms a spark across a pair of electrodes and causes a rapid rise of temperature. This high temperature generates an oscillation of gas bubbles that produces sound (Nedwell and Howell, 2004). Vessel traffic will increase in the vicinity of a wind farm before its construction and continue through to decommissioning.

The installation of wind turbine foundations is one of the most significant activities during construction. Drilling takes place in conjunction with pile driving because hard rocky substrates require a drilled socket for the pile. Dredging takes place in order to modify the bathymetry of the wind farm site; a moored ship uses suction pipes to collect sand and gravel from the seafloor, and then discharge pipes to deposit the sand and gravel at the discharge site (Nedwell and Howell, 2004). In order to protect cables from damage, they are trenched into the seafloor; however, trenching can only take place in seabed formed from soft sediment (Nedwell and Howell, 2004). The final stage of a wind farm's lifecycle is its decommissioning, the majority of which will be a reflection of the installation process. Noise data regarding the above-mentioned works are summarized in Table 1. The necessary terms used in Table 1 will be explained. The A-weighting system (dB(A)) is used to discriminate between low and high frequency extremes to approximate the auditory sensitivity of the human ear (review in Gamboa and Munda (2007)). For example, acoustic discomfort for human hearing ranges from 55 to 65 dB(A). The same concept can be applied to marine species. In the dBht (species), the suffix "ht" indicates that a sound is weighted by a species' hearing threshold (their "ht"); that is, the level expressed in this scale is species-specific and corresponds to each species' sound perception (Nedwell and Howell, 2004).

When a hammer strikes a pile, sound is generated in the air surrounding both the hammer and the pile. This sound may enter the water by an airborne path where airborne noise is incident upon the air–water interface and a portion of the sound energy enters the water (Nedwell and Howell, 2004). In sum, there are three noise paths: an airborne path, a waterborne path, and a ground-borne path.

Table 1. Underwater noise levels during pre-construction and construction of offshore wind farms [adapted from Nedwell and Howell (2004)]

Noise source	Measurement unit	Sound level at source point	
		Min.	Max.
Shipping & machinery	Linear SPL	152	192
	dBht (*Salmo salar*)	30	60
Geophysical survey	Linear peal SPL	215	260
	Peak dBht (*Gadus morhua*)	142	195
	Peak dBht (*Orcinus orca*)	159	216
	Peak dBht (*Phocoena phocoena*)	144	203
Drilling	Linear SPL	145	191
Dredging	Peak spectral energy (80-200 Hz)	177	
Trenching	dBht (*Orcinus orca*)	150	
	dBht (*Phocoena phocoena*)	120	
	dBht (*Phoca vitulina*)	100	
	dBht (*Gadus morhua*)	75	
	dBht (*Limanda limanda*)	50	
	dBht (*Salmo salar*)	50	

SPL = sound pressure level (underwater dB)
dBht = species-specific threshold sound level

3.3. Operation Noise

Air movement over an offshore wind turbine's entire structure, including the turbine blades and the hydrodynamic forces from passing waves, induces structural vibrations.

Figure 3 shows several paths (Betke et al., 2004; Nedwell and Howell, 2004): (i) the rotor blades generate aerodynamic noise. As this noise passes through the air, it may enter the water by an airborne path; (ii) airborne noise is created by vibrations in a turbine support (tower), a nacelle, and rotor blades, so a portion of this airborne noise encounters the air-water interface; (iii) a structure foundation is in contact with the seabed, and structural vibrations induce waves that can travel through the seabed. These waves propagate sound that tends to leak upwards into the water, and (iv) structural vibrations are transmitted to the lower part of the tower and encounter the turbine-water interface. This path contributes the most underwater noise (Nedwell and Howell, 2004).

Pre-construction, construction, and decommissioning take place over a period of months, but operational noise will occur over a number of years. The measurements of underwater noise from operating turbines at Utgrunden offshore wind farm in Sweden have been reported (Betke et al., 2004). The measured noise profiles are shown in Figure 4. In this offshore wind farm, the generator runs at 1100 rpm when the wind speed is low, but its operating rotation rises rapidly to the nominal value of 1800 rpm. Therefore, two noise profiles are illustrated in Figure 4 – one for low wind speeds and one for moderate and strong wind. Based on the data shown in Figure 4, underwater noise from wind turbines is of a low level and a low frequency; however, the level slightly increases as the wind speed increases, and it clearly increases as the installation capacity increases.

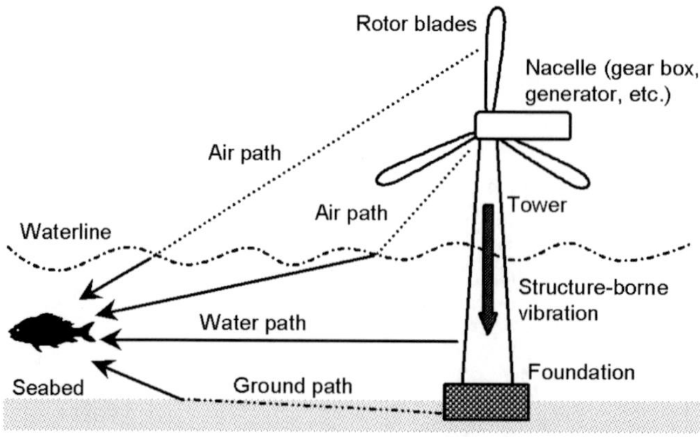

Figure 3. Transmission mechanism of underwater noise from an offshore wind turbine under operation (redrawn from Betke et al. (2004), and Nedwell and Howell (2004)).

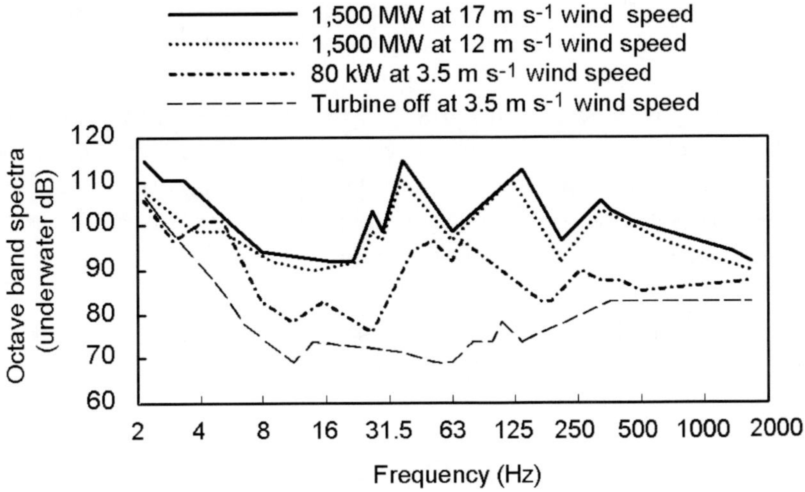

Figure 4. Underwater noise (1/3 octave band spectra) recorded 110 m away from the wind turbine in different turbine states (redrawn from Betke et al. (2004)).

CONCLUSION

Although there are concerns raised by the increased presence of anthropogenic sound in the aquatic environment, very little is known about the effects of exposure to such sounds on marine life, and far less is known about the effects on fishes. As for underwater noise induced by offshore wind farms, only a few studies deal with its impact on fish. Considering the description presented thus far, it is possible to hypothesize that noise from offshore wind farms may have ecological impacts on fish: (i) the short-term potential impact during pre-construction (cf. Table 1); (ii) the short-term intensive impact during construction (cf Table 1); and (iii) the physiological and/or masking effects that may occur over a long period while the wind farm is in operation (cf. Figure 4). Nevertheless, there are contradictions concerning fish responses to the noise of offshore wind farms; for example, a Swedish report (offshore wind power project in Norgersund) has shown that the 220-kW turbine has no negative impact on fish (review in European Thematic Network on Wave Energy (2003)). That is, there are significant data gaps. It may be necessary for a detailed study on the impact of underwater noise from offshore wind farms on marine life to be compiled in a systematic manner.

The discussed subject (underwater noise) has its roots in carbon offsetting. The European Union (EU) has a binding target of 20% for renewable energy in its energy mix by 2020. This EU target has been boosting the proportion of wind power in the energy sector. The penetration of wind power should be contemplated not only from a specific viewpoint (GHG reduction), but also from a general outlook because we must defend an environmental principle: processes that solve one environmental problem but cause another should be avoided. Since the noise effects of offshore wind farms on marine species are potentially adverse, it is suggested that offshore wind farms should be listed in the noise databases at least in order to promote rational environment management.

ACKNOWLEDGMENTS

Parts of this work are supported by Centro de Estudos de Recursos Naturais, Ambiente e Sociedade (CERNAS). The author is grateful to Ms. C. Lentfer for English review.

REFERENCES

Betke, K., Glahn, M. and Matuschek, R. (2004). Underwater noise emission from offshore wind turbines. In: *Proceedings of the Joint French–German Meeting CFA/DAGA – Congres Francais d'Acoustique and Deutsche Jahrestagung fur Akustik* (pp. 591-592). Strasbourg..

Division for Ocean Affairs and the Law of the Sea (2003). *Report of the Secretary-General: Oceans and the Law of the Sea*. Document No. A/58/65. Washington, DC: The United Nations Office of Legal Affairs.

Edge, G. and Blanchard, L. (2007). *Delivering Offshore Wind Power in Europ*e. Brussels: European Wind Energy Association.

Engas, A., Lokkeborg, S., Ona, E. and Soldal, A. V. (1996). Effects of seismic shooting on local abundance and catch rates of cod (*Gadus morhua*) and haddock (*Melanogrammus aeglefinus*). *Canadian Journal of Fisheries and Aquatic Sciences*, 53, 2238-2249.

European Thematic Network on Wave Energy (2003). *Wavenet Report,* No. ERK5 CT 1999/2001. Brussels: Research Directorate-General of the European Commission.

Gamboa, G. and Munda, G. (2007). The problem of wind farm location – a social multicriteria evaluation framework. *Energy Policy*, 35(3), 1564-1583.

Harrison, P. (2014). *The noise pollution ecosystem*. Shreveport: Sound Fighter Systems.

Hastings, M. C. and Popper, A. N. (2005). *Effects of Sound on Fish*. Report No. 43A 0139. Sacramento: California Department of Transportation.

Hastings, M. C. (2008). Coming to terms with the effects of ocean noise on marine animals. *Acoustics Today,* 4, 22-34.

Hubbs-Sea World Research Institute (2007). *Aquatic Noise Pollution from Oil Tankers and Escort Vessels in Prince William Sound, its Effects and Impacts on the Marine Environment of the Sound*. Report No. 854 071. Anchorage: Prince William Sound Regional Advisory Council.

Johnson, J. S. (2003). *SURTASS LFA environmental compliance experience*. Presentation at Environmental Consequences of Underwater Sound, May 12-16, San Antonio (TX).

McCauley, R. D., Fewtrewll, J. and Popper, A. N. (2003). High intensity anthropogenic sound damages fish ears. *Journal of the Acoustical Society of America,* 113(1), 638-642.

McDonald, M. A., Hildebrand, J. A. and Wiggins, S. M. (2006). Increases in deep ocean ambient noise in the Northeast Pacific west of San Nicolas Island, California. *Journal of the Acoustical Society of America*, 120(2), 711-718.

National Research Council (1994). *Low-Frequency Sound and Marine Mammals: Current Knowledge and Research Need*. Washington DC: National Academy Press.

National Research Council (2000). *Marine Mammals and Low-Frequency Sound: Progress Since 1994*. Washington DC: National Academy Press.

National Research Council (2003). *Ocean Noise and Marine Mammals*. Washington DC: National Academy Press.

National Research Council (2005). *Marine Mammal Populations and Ocean Noise: Determining When Noise Causes Biologically Significant Effects*. Washington DC: National Academy Press.

Nedwell, J. R. and Howell, D. (2004). *A Review of Offshore Wind Farm-related Underwater Noise Sources*, Report No. 544 R 0308. London: Collaborative Offshore Wind Energy Research into the Environment.

Nieukirk, S. L., Stafford, K. M., Mellinger, D. K., Dziak, R. P. and Fox, C. G. (2004). Low-frequency whale and seismic airgun sounds recorded in the mid-Atlantic Ocean. *Journal of the Acoustical Society of America*, 115(4), 1832-1843.

Office of Marine Programs (2007). *Science of Sound in the Sea*. Kingston: University of Rhode Island.

Pauly, D. (2003). Ecosystem impacts of the world's marine fisheries. *Global Change Newsletter*, 55, 21.

Popper, A. N., Fay, R. R., Platt, C. and Sand, O. (2003). Sound detection mechanisms and capabilities of teleost fishes. In: Colin, S. P. and Marshall, N. J. (Eds.), *Sensory Processing in Aquatic Environments* (pp. 3-38). New York: Springer.

Popper, A. N. and Hastings, M. C. (2009). The effects of anthropogenic sources of sound on fishes. *Journal of Fish Biology*, 75, 455-489.

Reeves, A. (2003). *Wind Energy for Electric Power – A REPP Issue Brief.* Washington DC: Renewable Energy Policy Project.

Simmond, M. P. and Dolman, S. J. (2008). All at sea – renewable energy production in the context of marine nature conservation. In: Evans, P. G. H. (Ed.), *Proceedings of the Workshop on Offshore Wind Farms and Mammals – Impact and Methodologies for Assessing Impacts* (pp. 6-11), ECS Special Publication Series No. 4. Oxford: European Cetacean Society.

Smith, M. E., Kane, A. S. and Popper, A. N. (2004a). Acoustical stress and hearing sensitivity in fishes: does the linear threshold shift hypothesis hold water? *Journal of Experimental Biology*, 207(20), 3591-3602.

Smith, M. E., Kane, A. S. and Popper, A. N. (2004b). Noise-induced stress response and hearing loss in goldfish (*Carassius auratus*). *Journal of Experimental Biology*, 207(3), 427-435.

Southall, B. L., Bowles, A. E., Ellison, W. T., Finneran, J. J., Gentry, R. L., Greene C. R. Jr, Kastak, D., Ketten, D. R., Miller, J. H., Nachtigall, P. E., Richardson, W. J., Thomas, J. A. and Tyack, P. L. (2007). Marine mammal noise exposure criteria: Initial scientific recommendations. *Aquatic Mammals,* 33**,** 411-521.

Slotte, A., Kansen, K., Dalen, J. and Ona, E. (2004). Acoustic mapping of pelagic fish distribution and abundance in relation to a seismic shooting area off the Norwegian west coast. *Fish Research*, 67, 143-150.

Tyack, P. L, and Miller, E. H. (2002). Vocal anatomy, acoustic communication and echolocation. In: A. R. Hoelzel (ed.), *Marine mammal biology* (pp. 142-184). Oxford: Blackwell Science.

Wahlberg, M. and Westerberg, H. (2005). Hearing in fish and their reactions to sounds from offshore wind farms. *Marine Ecology Progress Series*, 288, 295-309.

Weilgart, L. S. (2008). *The impact of ocean noise pollution on marine biodiversity*. Washington DC: Animal Welfare Institute.

Wenz, G. M. (1962). Acoustic ambient noise in the ocean: spectra and sources. *Journal of the Acoustical Society of America,* 34, 1936-1956.

In: Noise Pollution
Editor: Eugene Roberto Nicchi
ISBN: 978-1-63321-110-0

Chapter 8

AN EXPERIMENTAL STUDY OF A ONE-CHAMBER AND TWO-CHAMBER MODIFIED HELMHOLTZ RESONATORS

Min-Chie Chiu *

*Department of Mechanical and Automation Engineering, Chung Chou University of Science and Technology, Taiwan, ROC

ABSTRACT

Noise control is essential in an enclosed machine room where high levels of reverberant and direct sound are encountered. Because pure tone noise is harmful to workers and may lead to psychological and physiological ailments, a traditional Helmholtz resonator that reduces pure tone noise has been commonly employed. However, the conventional Helmholtz resonator which is equipped with an externally extended tube requires a larger space. For existing plants that are in need of sufficient operation and maintenance space, a smaller designed Helmholtz resonator is necessary. In order to overcome the above drawbacks of an oversized Helmholtz resonator, a modified Helmholtz resonator that is compact and equipped with an internally extended tube is proposed. To appreciate the response of resonance frequencies and

* Corresponding author: Min-Chie Chiu, Department of Mechanical and Automation Engineering, Chung Chou University of Science and Technology, No. 6, Lane 2, Sec.3, Shanchiao Rd., Yuanlin, Changhua 51003, Taiwan, R.O.C. E-mail: minchie.chiu @msa.hinet.net

sound absorption coefficients with respect to various acoustical mechanisms and design parameters, an experimental study that varies the design parameters (L_k, d_k, F_k, H) is performed. Additionally, a two-layer resonator connected by two one-layer resonators in series is also discussed.

Results reveal that the equivalent shape factor (F_k) will increase when the ratio of the length to diameter for the resonating tube decreases. Both the resonance frequency (f_{res}) and related sound absorption coefficient (*abs*) will increase when the d_k increases. In addition, the resonance frequency (f_{res}) will decrease and the related sound absorption coefficient (*abs*) will increase when the H increases. Subsequently, both the resonance frequency (f_{res}) and sound absorption coefficient (*abs*) will decrease simultaneously when the L_k increases.

Consequently, the resonance frequency (f_{res}) of the two-layer modified Helmholtz resonator (2MHR-001 and 2MHR-002) will be tuned at higher frequencies than that of the one-layer resonator (1MHR030 and 1MHR040).

Keywords: Helmholtz; resonator; resonating tube, air space

1. Nomenclature

abs	normal sound absorption coefficient (= α)
C_{res}	the acoustic compliance
d_k	the diameter of a resonating tube in a Helmholtz resonator
F_k	the shape factor for a resonating tube in a modified Helmholtz resonator
f_{res}	the resonance frequency for a Helmholtz resonator
H_i	the depth of the air space at the i-th layer Helmholtz resonator
L_k	the length of a resonating tube for a Helmholtz resonator
L_{res}	the acoustic inertia
R_{res}	the acoustic resistance

S_k the cross section area of a resonating tube in a Helmholtz resonator

V_c the resonating volume of a Helmholtz resonator

2. Introduction

Pure tone noise emitted from transformers and induced fans is harmful to people [1]. However, the energy of a low frequency pure tone is difficult to eliminate. And, because equipment is often located within an enclosed machine room, the noise will be amplified due to reverberation. In addition, this integrated exposed noise will exceed the standards set by OSHA [2]. Therefore, the development of an acoustical panel hybridized with a resonator used to eliminate pure tones is necessary. Research in acoustical panels has, of course, been well documented. However, most of the research focused on the analysis of sound absorbing material [3, 4, 5] and wide-band acoustical panels [6, 7, 8, 9]. Previous studies [10, 11, 12, 13, 14, 15] developed multi-layer sound absorbers with wide-band sound absorption ability. Yet, the sound absorption ability for dealing with lower frequency pure tones was insufficient. A traditional Helmholtz resonator [16] provides sufficient sound absorption for pure tones. But, the Helmholtz resonator with an externally extended tube is bulky and will interfere with operations and maintenance inside the machine room. Therefore, a compact modified Helmholtz resonator with an internally extended tube is proposed. To appreciate the sound absorption ability with respect to the related resonance frequencies, an experimental study that varies the design parameters (L_k, d_k, F_k, H) is carried out. Moreover, a two-layer resonator that is connected by two one-layer resonators in series will also be discussed.

3. Mathematical Models

For a modified Helmholtz resonator (*MHR*) with an internally extended tube shown in Figure 1, the acoustical impedance (Z) of the modified Helmholtz resonator (*MHR*) is

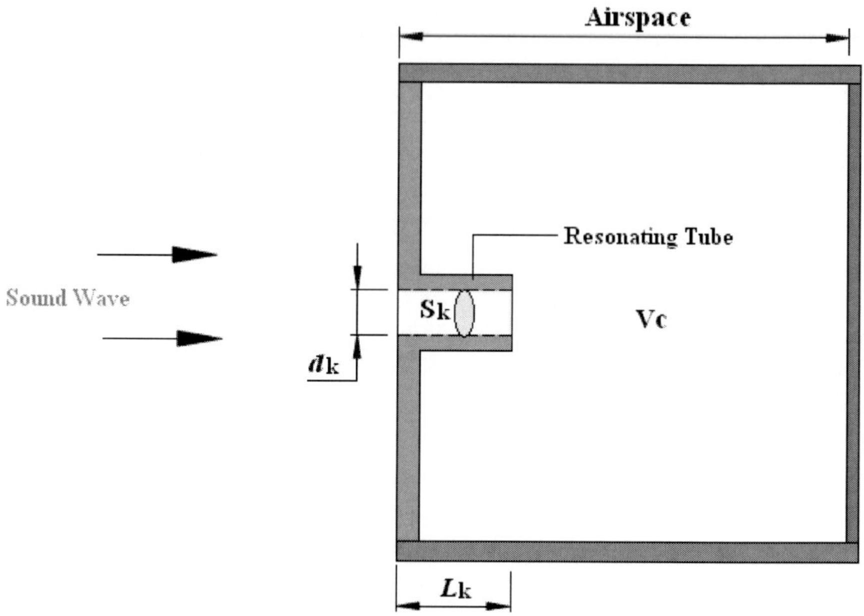

Figure 1. A one-layer modified Helmholtz Resonator.

$$Z = R_{res} + j\omega L_{res} + \frac{1}{j\omega C_{res}} \tag{1}$$

$$L_{res} = \frac{\rho_o (F_k L_k)}{S_k}; C_{res} = \frac{V_c}{\rho_o c_o^2}; R_{res} = \frac{\rho_o \omega^2}{2\pi c_o} \tag{2}$$

where R_{res} is the acoustic resistance, L_{res} is the acoustic inertia, C_{res} is the acoustic compliance), L_k is the length of the internally extended tube, S_k is the cross section area of the extended tube, V_c is the volume of the *MHR*, F_k is the shape factor for a resonating tube inside the *MHR*.

Neglecting the acoustic resistance, the resonance frequency (f_{res}) is derived as

$$f_{res} = \frac{c_o}{2\pi}\sqrt{\frac{S_k}{V_c(F_k L_k)}} \quad (3)$$

To appreciate the actual shape factor (F_k) and the resonance frequency (f_{res}) for various *MHR*s equipped with various design parameters (the diameter of the resonating tube(d_k), the length of the resonating tube (L_k), and the cavity of the *MHR* (V_c)), an experimental study using a device with two-microphone impedance tubes shown in Figure 2 is carried out. Moreover, the acoustical performance of the one-layer *MHR* and the two-layer *MHR* will also be tested and compared.

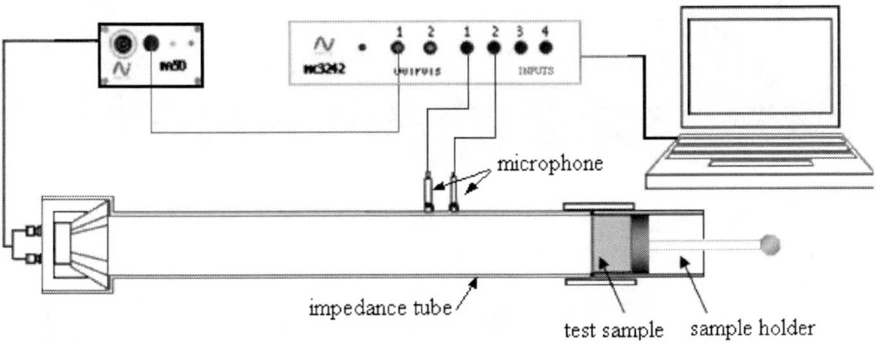

Figure 2. An experimental setup for the two-microphone impedance tube.

4. Case Studies

A one-layer modified Helmholtz resonator (*1MHR*) with various design parameters — the depth of air space (H), the diameter of the resonating tube (d_k), and the length of the resonating tube (L_k) shown in Figure 3 — is investigated. The ranges of the parameters are (H=50mm, 100mm, and 150mm), (d_k=5mm, 10mm, 15mm, and 20mm), and (L_k=25mm, 50mm, 75mm, and 100mm). Moreover, the acoustical performances for two kinds of double-layer Helmholtz resonators — a *2MHR-001* (1st Layer: d_k=15mm L_k=25mm H_1=50mm + 2nd Layer: 5T d_k=20mm L_k=25mm H_2=50mm); and the other, a *2MHR-002* (1st Layer: 5T d_k=20mm L_k=25mm H_1=50mm + 2nd Layer: 10T d_k=15mm L_k=25mm H_2=50mm) — are also experimentally analyzed.

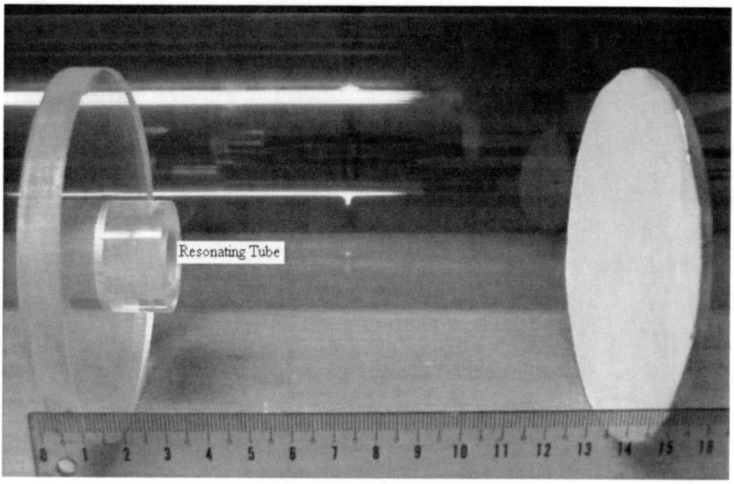

Figure 3. A one-layer modified Helmholtz resonator (*1MHR*).

5. Experimental Setup

As indicated in Figure 2, an acoustical impedance tube with two microphones that are linked to a FFT (Fast Fourier Transform) is established. The resonance frequencies of the one-layer modified Helmholtz resonator (*1MHR*) and the double-layer Helmholtz resonators (*2MHR-001* and *2MHR-002*) will be detected by using the two-microphone acoustical impedance tube.

6. Results and Discussion

6.1. Results

6.1.1. A One-Layer Modified Helmholtz Resonator (1MHR)

For a one-layer modified Helmholtz resonator shown in Figure 4, the experimental results of the sound absorption coefficient and related resonant frequencies for various ranges of design parameters (H, d_k, L_k) are summarized and depicted in Table 1. On the basis of a fixed L_k (=25mm) and H(=50mm), the sound absorption coefficient with respect to various d_k is shown in Figure 5. Similarly, at the values of L_k (=25mm) and H(=100mm), the sound absorption coefficient with respect to various d_k is plotted in Figure 6.

Table 1. Peak frequency and sound absorption with respect to various single-layer modified Helmholtz resonators

item	H(mm)	d_k (mm)	L_k (mm)	f_{res} (Hz)	α at f_{res}	F_k
1MHR-010	50	5	25	100	0.44	0.149
1MHR-020	50	10	25	175	0.69	0.1946
1MHR-030	50	15	25	250	0.90	0.2145
1MHR-040	50	20	25	280	0.99	0.304
1MHR-110	100	5	25	90	0.55	0.0919
1MHR-120	100	10	25	118	0.85	0.214
1MHR-130	100	15	25	160	0.99	0.2619
1MHR-220	150	10	25	90	0.92	0.245
1MHR-121	100	10	50	118	0.79	0.214
1MHR-221	150	10	50	90	0.84	0.184
1MHR-122	100	10	75	67	0.52	0.088
1MHR-222	150	10	75	106	0.70	0.147
1MHR-213	150	5	100	56	0.61	0.039
1MHR-223	150	10	100	80	0.76	0.077

To appreciate the influence of the resonance frequency (f_{res}) with respect to the air space (H), the profile of sound absorption coefficient with respect to various H at a fixed L_k (=25mm) and d_k (=10mm) is plotted in Figure 7.

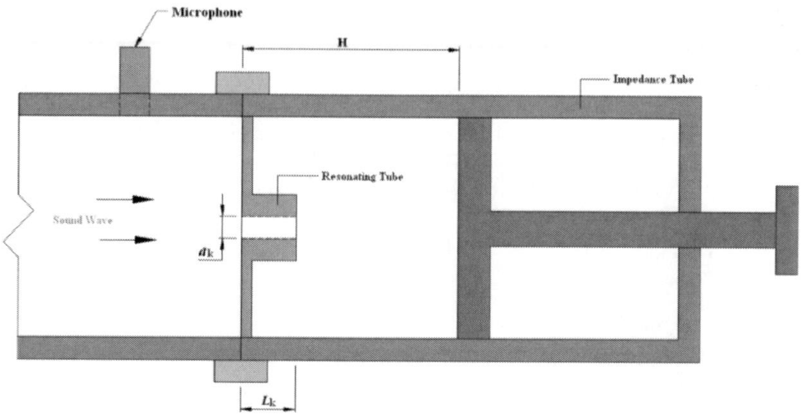

Figure 4. A single layer modified Helmholtz resonator.

Similarly, the sound absorption coefficient with respect to various *H* at a fixed L_k(=50mm) and d_k(=10mm) is plotted in Figure 8. In addition, the sound absorption coefficient with respect to various *H* at a fixed L_k(=75mm) and d_k(=10mm) is depicted in Figure 9. Moreover, the sound absorption coefficient with respect to various d_k at a fixed L_k(=100mm) and *H* (=150mm) is illustrated in Figure 10. Considering the influence of the resonance frequency (f_{res}) with respect to the air space (*H*), the sound absorption coefficient with respect to various L_k at a fixed *H* (=100mm) and d_k(=10mm) is depicted in Figure 11.

6.1.2. A Two-Layer Modified Helmholtz Resonator (2MHR)

As indicated in Figure 12, experimental work on a two-layer modified Helmholtz resonator is performed. The experimental results of the two-layer modified Helmholtz resonator are shown in Table 2. Two kinds of two-layer modified Helmholtz resonators (*2MHR-001* and *2MHR-002*) are proposed. Here, the first one (2MHR-001) is composed of 1MHR-030 (first layer resonator) and 1MHR-040 (second layer resonator). Also, the second one (2MHR-002) is composed of 1MHR-040 (first layer resonator) and 1MHR-030 (second layer resonator). Results reveal that the profile of sound absorption coefficients with respect to 1MHR-030, 1MHR-040, and 2MHR-001 is plotted in Figure 13. Moreover, the sound absorption coefficient for 1MHR-030, 1MHR-040, and 2MHR-002 is plotted in Figure 14.

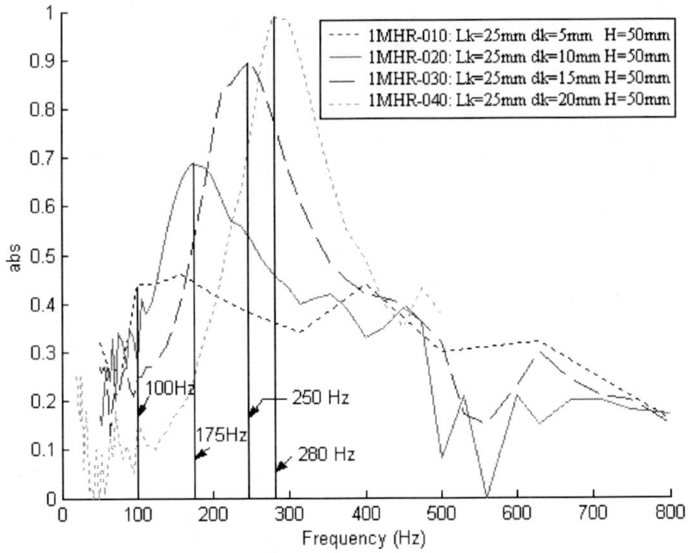

Figure 5. Sound absorption coefficient with respect to one-layer modified Helmholtz resonators (*H* =50mm) with various diameters of resonating tubes (1MHR-010, 1MHR-020, 1MHR-030, 1MHR-040).

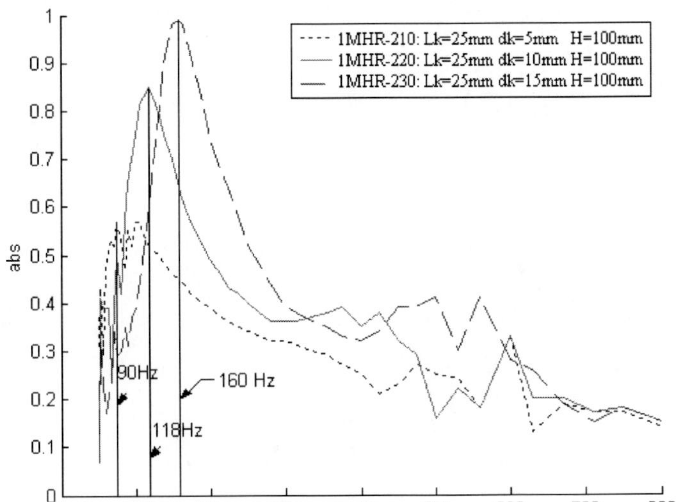

Figure 6. Sound absorption coefficient with respect to one-layer modified Helmholtz resonators (*H* =100mm) with various diameters of resonating tubes (1MHR-210, 1MHR-220, 1MHR-230).

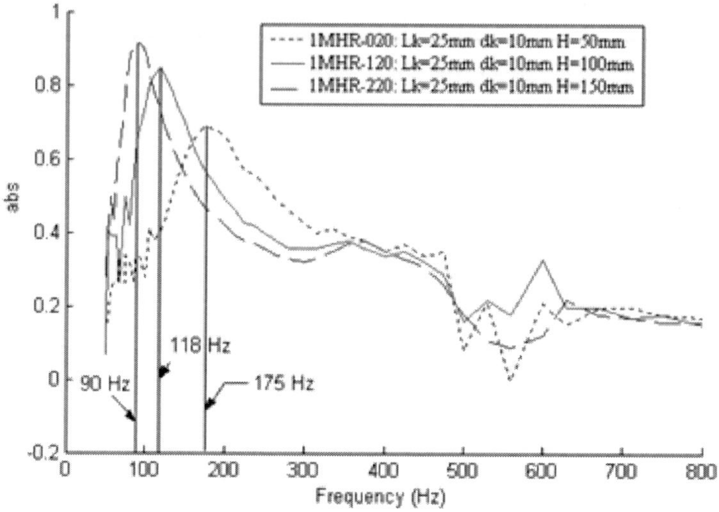

Figure 7. Sound absorption coefficient with respect to one-layer modified Helmholtz resonators (d_k =10mm, L_k =25mm) with various air space(H) (1MHR-020, 1MHR-120, 1MHR-220).

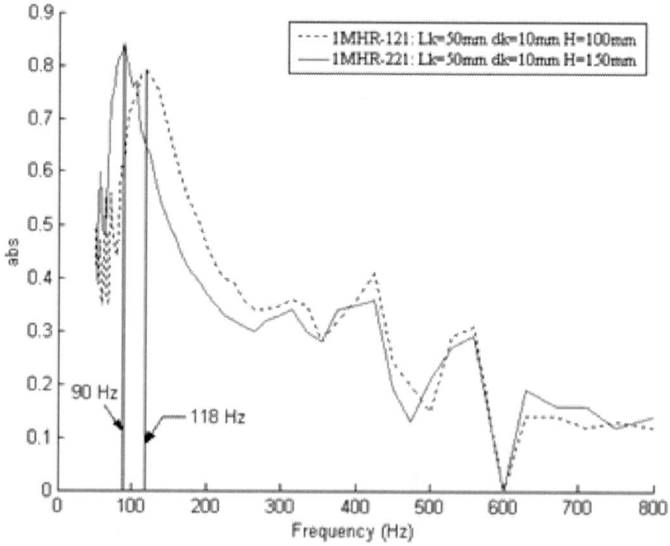

Figure 8. Sound absorption coefficient with respect to one-layer modified Helmholtz resonators (d_k =10mm, L_k =50mm) with various air space(H) (1MHR-121, 1MHR-221).

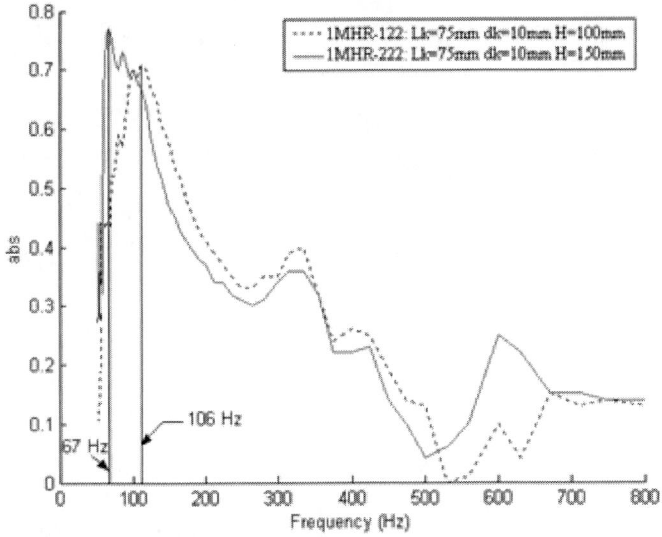

Figure 9. Sound absorption coefficient with respect to one-layer modified Helmholtz resonators (d_k =10mm, L_k =75mm) with various air space(*H*) (1MHR-122, 1MHR-222).

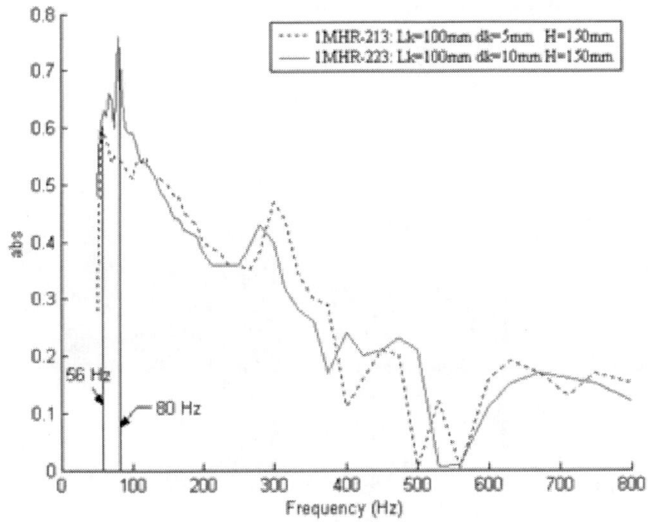

Figure 10. Sound absorption coefficient with respect to one-layer modified Helmholtz resonators (*H* =150mm, L_k =100mm) with various diameters of resonating tubes (d_k) (1MHR-213, 1MHR-223).

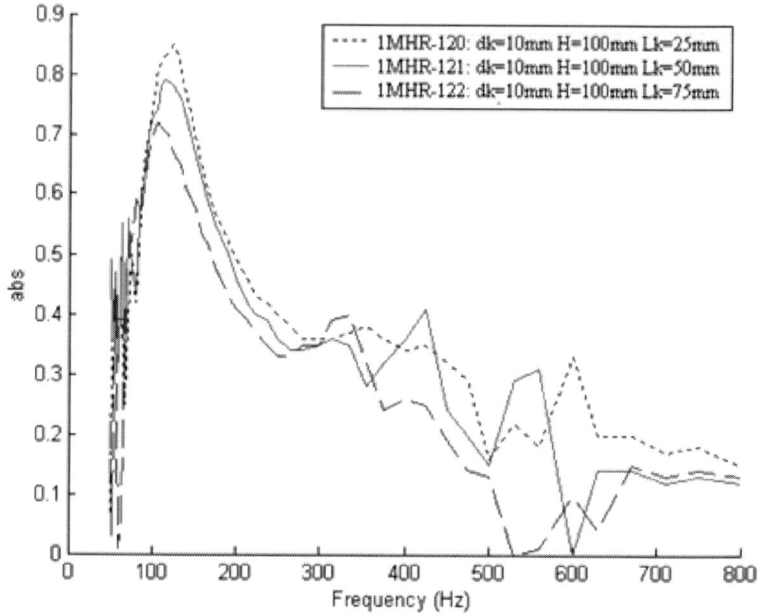

Figure 11. Sound absorption coefficient with respect to one-layer modified Helmholtz resonators (d_k =10mm, H =150mm) with various lengths of resonating tubes (L_k) (1MHR-120, 1MHR-121, 1MHR-122).

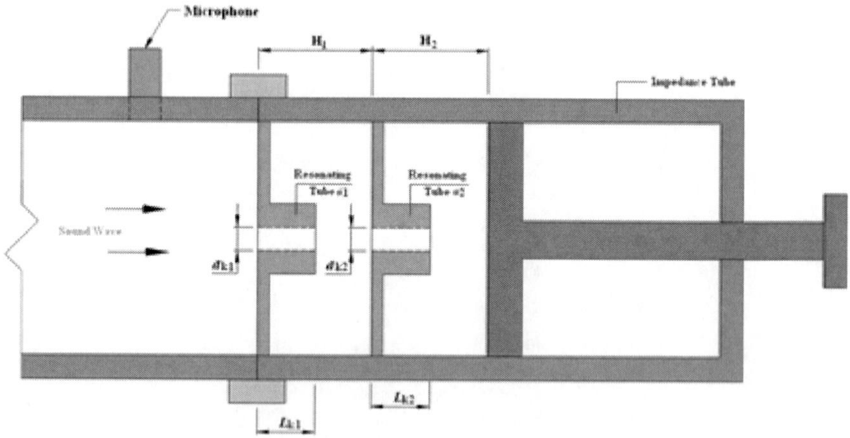

Figure 12. A double-layer modified Helmholtz resonator.

Table 2. Peak frequency and sound absorption with respect to various double-layer modified Helmholtz resonators

item	First Layer			Second Layer			Result			
	H_1(mm)	d_{k1} (mm)	L_{k1} (mm)	H_2(mm)	d_{k2} (mm)	L_{k2} (mm)	f_{res-1} (Hz)	α at f_{res-1}	f_{res-1} (Hz)	α at f_{res-1}
2MHR-001	50	15	25	50	20	25	140	0. 98		
2MHR-002	50	20	25	50	15	25	160	0.98	375	0.79

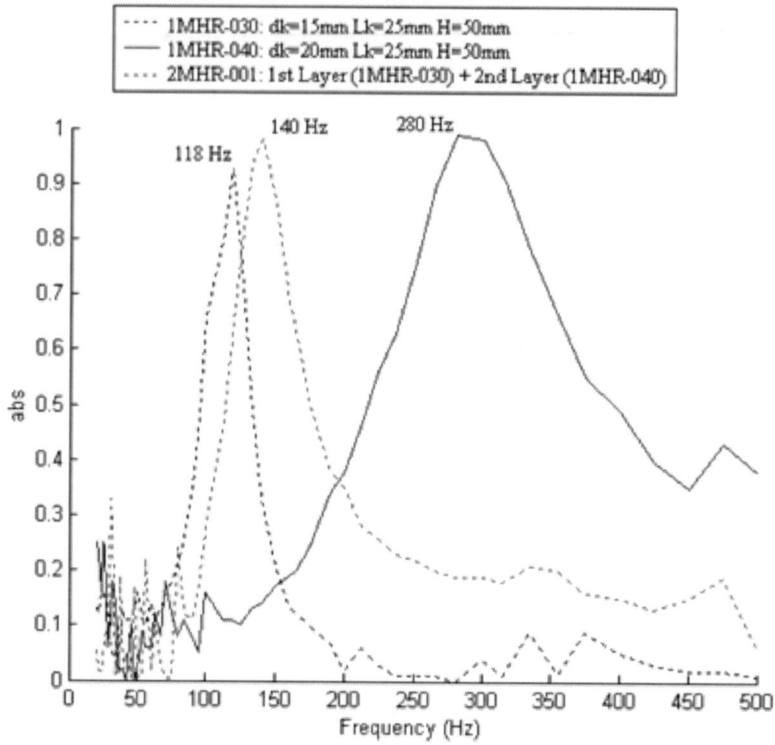

Figure 13. Sound absorption coefficient with respect to two one-layer modified Helmholtz resonators (1MHR-030 and 1MHR-040) and a double–layer modified Helmholtz resonator (2MHR-001).

6.2. Discussion

As indicated in Table 1, on the basis of the same air space (H=50mm) and resonating tube length (L_k=25mm), the equivalent shape factor (F_k) will increase from 0.149 to 0.304 when the diameter of the resonating tube increases from 5mm to 20mm. Similarly, at a fixed H(=50mm) and L_k(=25mm), the equivalent shape factor (F_k) will increase from 0.0919 to 0.2619 when the diameter of the resonating tube increases from 5mm to 15mm. Obviously, on the basis of same H and L_k, the equivalent shape factor (F_k) will increase when the ratio of length to diameter for the resonating tube decreases.

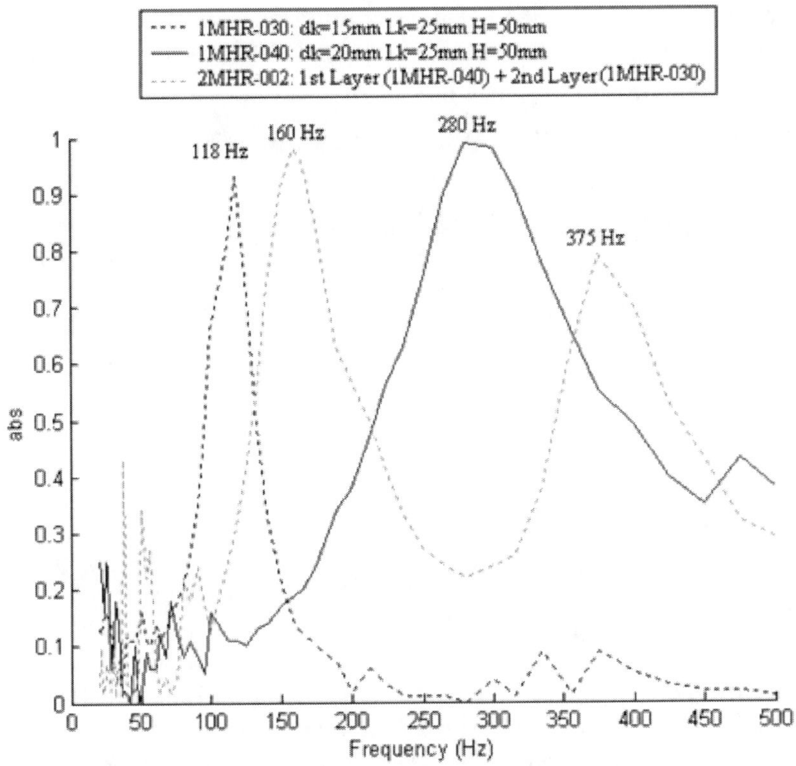

Figure 14. Sound absorption coefficient with respect to two one-layer modified Helmholtz resonators (1MHR-030 and 1MHR-040) and a double–layer modified Helmholtz resonator (2MHR-002).

As indicated in Eq. (3), the resonance frequency (f_{res}) will be proportional to the diameter of the resonating tube (d_k) and inversely proportional to the shape factor (F_k), the tube length (L_k), and the air space (H). The experimental results shown in Figures 5, 6, and 10 indicate that on the basis of same L_k and H, both the resonance frequency (f_{res}) and its sound absorption coefficient (*abs*) will increase when the d_k increases.

Likewise, as indicated in Figures 7, 8, and 9, on the basis of same L_k and d_k, the resonance frequency (f_{res}) will decrease, and its sound absorption coefficient (*abs*) will increase when the H increases. Moreover, as indicated in Figure 11, on the basis of same H and d_k, both the resonance frequency (f_{res}) and its sound absorption coefficient (*abs*) will decrease when the L_k increases.

The double–layer modified Helmholtz resonator 2MHR-001 is organized with 1MHR030 (the first layer) in series with 1MHR040 (the second layer). As indicated in Figure 13, the individual resonance frequency (f_{res}) of 1MHR030 and 1MHR040 are 118 Hz and 280 Hz. The resultant resonance frequency (f_{res}) of 2MHR-001 is 140 Hz. Similarly, for the double–layer modified Helmholtz resonator 2MHR-002 that is organized with 1MHR040 (the first layer) in series with 1MHR030 (the second layer), the resultant resonance frequencies (f_{res}) of 2MHR-001 shown in Figure 14 are 160 Hz and 375 Hz. Obviously, the resultant resonance frequency (f_{res}) of the two-layer modified Helmholtz resonator (2MHR-001 and 2MHR-002) is higher than the individual single layer resonator (1MHR030 and 1MHR040).

CONCLUSION

Considering the necessity of maintenance and operation in a work shop, the compactness of a Helmholtz resonator installed inside the room is compulsory. To shorten the volume of the resonator, a Helmholtz resonator equipped with an externally extended tube is replaced by an internally extended tube. To understand the acoustical influence of design parameters (L_k, d_k, F_k, and H) in the modified Helmholtz resonator, a series of experiments was carried out. Results revealed that on the basis of the same H and L_k, the equivalent shape factor (F_k) will increase when the ratio of length to diameter for the resonating tube decreases. In addition, at the same L_k and H, both the resonance frequency (f_{res}) and the related sound absorption coefficient (*abs*) will increase when the d_k increases. Similarly, with the same L_k and d_k, the resonant frequency (f_{res}) will decrease and its sound absorption coefficient (*abs*) will also increase if H increases. Moreover, based on the same H and d_k, both the resonant frequency (f_{res}) and sound absorption coefficient (*abs*) will decrease when L_k increases. Consequently, for the modified Helmholtz resonator equipped with a two-layer chamber, the resonant frequency (f_{res}) of the two-layer modified Helmholtz resonator (2MHR-001 and 2MHR-002) will be higher than that of the individual one-layer resonator (1MHR030 and 1MHR040).

ACKNOWLEDGEMENTS

The author acknowledges the financial support of the National Science Council (NSC100-2622-E-235-001-CC3), ROC.

REFERENCES

[1] P. N. Cheremisinoff, P. P. Cheremisinoff, Industrial Noise Control Handbook, Ann Arbor Science, Michigan, 1977.

[2] B. C. Alley, R. M. Dufresne, N. Kanji, M. R. Reesal, Costs of workers' compensation claims for hearing loss, *Journal of Occupational Medicine*. (1989), 31, pp. 134-138.

[3] M. E. Delany, E. N. Bazley, Acoustical properties of fibrous absorbent materials, *Applied Acoustics*. (1969), 13, pp. 105-116.

[4] D. A. Bies, C. H. Hansen, Engineering Noise Control, Unwin Hyman, New York, 1988.

[5] L. L. Beranek, I. L. Ver, Noise and Vibration Control Engineering, John Wiley & Sons, New York, 1992.

[6] D. Y. Maa, Microperforated-panel wideband absorbers, *Noise Control Engineering Journal*. (1987), 29, pp. 77-84.

[7] L. Jinkyo, G. Swenson, Compact sound absorbers for low frequencies, *J. Noise Control Engineering Journal*. (1992), 38, pp. 109-117.

[8] F. C. Lee, W. H. Chen, Acoustic transmission analysis of multi-layer absorbers, J*ournal of Sound and Vibration*. (2001), 248, pp. 621-634.

[9] J. Liu, D.W. Herrin, Enhancing micro-perforated panel attenuation by partitioning the adjoining cavity,*Applied Acoustics*. (2010), 71, pp. 120-127.

[10] Y. C. Chang, L. J. Yeh, M. C. Chiu, G. J. Lai, Numerical optimization of sound absorption systems with constrained single-layer absorber, *Journal of Marine Science and Technology*. (2004), 12(2), pp. 86-92.

[11] Y. C. Chang, L. J. Yeh, M. C. Chiu, G. J. Lai, Shape optimization on constrained single-layer sound absorber by using GA method and mathematical gradient methods, *Journal of Sound and Vibration*. (2005), 286(4-5), pp. 941-961.

[12] Y. C. Chang, L. J. Yeh, M. C. Chiu, Optimization of composite absorbers on constrained sound reverberant system by using simulated annealing, *Applied Acoustics*. (2005), 66, pp.341-352.

[13] M. C. Chiu, L. J. Yeh, Y. C. Chang, G. J. Lai, Computer aided design on perforated single-layer absorbers under space constraints, *International Journal of Advanced Manufacturing Technology*. (2007), 32(5-6), pp. 537-546.

[14] Y. C. Chang, L. J. Yeh, M. C. Chiu, Optimization of constrained multi-layer absorbers by using genetic algorithms, *International Journal of Acoustics and Vibration*. (2004), 9(4), pp.175-185.

[15] Y. C. Chang, L. J. Yeh, M. C. Chiu, Optimization of double-layer absorbers on constrained sound absorption system by using genetic algorithm, *International Journal for Numerical Methods in Engineering*. (2005), 62, pp. 317-333.

[16] M.C. Junger, Helmholtz resonators in load-bearing walls, *Noise Control Engineering*. (1975), 4(1), pp. 17-25.

In: Noise Pollution
Editor: Eugene Roberto Nicchi
ISBN: 978-1-63321-110-0

Chapter 9

ROAD TRAFFIC NOISE EXPOSURE AND ANNOYANCE: A CROSS-SECTIONAL STUDY IN THE CITY OF SHARJAH, UAE

Hussein M. Elmehdi*
Applied Physics Department
University of Sharjah, Sharjah, United Arab Emirates

ABSTRACT

Previous research on noise levels arising form residential road traffic in the City of Sharjah, United Arab Emirates (UAE) has shown that noise levels in the city is high, exceeding national as well as international noise limits and standards. These studies have linked noise exposure in residential-industrial mixed area with annoyance, which is an indicator of more serious, chronic health conditions. In this chapter we report the results of a study conducted recently to explore the association between residential road traffic noise and self-reported annoyance for residents of the City of Sharjah. A comprehensive survey was conducted on randomly selected subjects aged 19-59 years, residing minimum of 5 years in the area and residing within 50 m of the roadways. The survey questions included two parts, the first of which contained demographic questions, while the second part contained questions aimed at gauging annoyance levels encountered by residents as a result of exposure to high noise levels. The results of the study showed that more than 67% of subjects

* Email: hmelmehdi@sharjah.ac.ae

feel that their neighborhood is noisy. To explore the association between noise levels and self-reported annoyance, noise levels were classified in units of 5 dB (A) from 65 dB (A) to 80 dB (A). Self-reported annoyance was marked at levels above 65-70 dB (A). Gender dependence was evident, where the association was found to be statistically significant for female subjects and the prevalence of annoyance was more for male subjects. Both age and period of residence were found to be significant predictors of annoyance. The results of this study further suggest the association residential noise exposure near roadways and self-reported annoyance, among the study subjects.

1. INTRODUCTION

Environmental noise is defined as the noise emitted from sources such as road, rail and air traffic, industries, construction and public works, and the neighborhood. Noise has been ranked high among forms of pollutions, which include air pollution, radioactive waste pollution, water pollutions, etc. In recent years, noise has received considerable worldwide attention as a result of the many studies linking noise pollution to various health effects that include auditory as well non-auditory health effects [1].It should be noted here that the effect of noise is seldom catastrophic, and is often only transitory. However, its adverse effects can be cumulative with prolonged or repeated exposure. Noise with daily activities and it causes sleep disruption, masking of speech, reduction in performance and the inability to enjoy one's property or leisure time. It impairs the quality of life. In addition, several studies have linked extended exposure to high noise levels to cardiovascular diseases, including high blood pressure and irregular heart beat [2]. It has also been reported that high noise levels have an economic effects, especially tourism and real estate sectors, where it has been reported that the prices of houses tend to be higher in quite areas [3]. There is also some evidence that noise can adversely affect general health and wellbeing in the same manner as chronic stress [4,5]. Because of these factors, and the recent public awareness of the adverse effects of noise in the last few decades, many studies and noise monitoring programs have been established in various parts in the world. These studies and programs focus on measurements of perceived sound levels in communities, in the workplace (occupational noise), near airports and in public parks. A second area of research, which is getting more attention in recent years, focuses on assessing the effects of noise on individuals and communities, i.e. noise annoyance.

In the United Arab Emirates (UAE), there is an increase in public awareness to noise pollution and its effects on individuals and the community. This is due to the attention that was generated after the publication of recent reports on noise levels in major UAE cities [6] and noise assessment near Dubai International Airport [7]. The results indicated that environmental noise levels, especially those due to transportation and aircraft noise, are high and well over the national as well as international standards. The current research has focused on the adverse effects of high noise levels the residents of UAE cities. In this paper we report the results of a comprehensive survey, which focuses on assessing the annoyance of citizens of the City of Sharjah (UAE) using social surveys. The aim is extend the study to cover other UAE cities such as Abu Dhabi, Dubai, and Alain.

2. Materials and Methods

The coastal city of Sharjah is the capital city of the Emirate of Sharjah. It is located in the northern part of the UAE on the coast of the Arabian Gulf (also known as the Persian Gulf). It is located at the eastern side of the Arabian Peninsula (see Figure 1). The city has witnessed over 50% increase in population in the last decade, reaching a population of approximately 800,000 according to the 2008 statistics [8]. This increase in population, and the economic prosperity led to over 500% increase in the number of vehicles on the streets, which is the main contributor to community noise in the city. This increase in motor vehicles is overwhelming the road grid, leading to major traffic jams on major roads as well as on residential streets, which is becoming a source of concern for environmentalists in the city. Couple the above with reckless driving, casual use of horns, high speed driving and huge volume of vehicles, especially those with heavy engine noises and the absence of sound proof fences, the noise pollution levels have exceeded international standards set by the WHO or NIHCDCNHL, as well as local standards. Noise pollution arising from such high noises has been recognized by officials and the public alike, as a major source that has behavioral, physiological and communicational effects, annoyance that is having its toll on individuals and on their performance and productivity [9,10].

The areas studied in this survey include four residential areas distributed throughout the City of Sharjah, including one is at the city center, and one in the industrial districts, both of which are known for heavy traffic and noisy streets. The other two are residential areas, which located at the outskirts of the city. The survey was handed out to people living in apartments and houses.

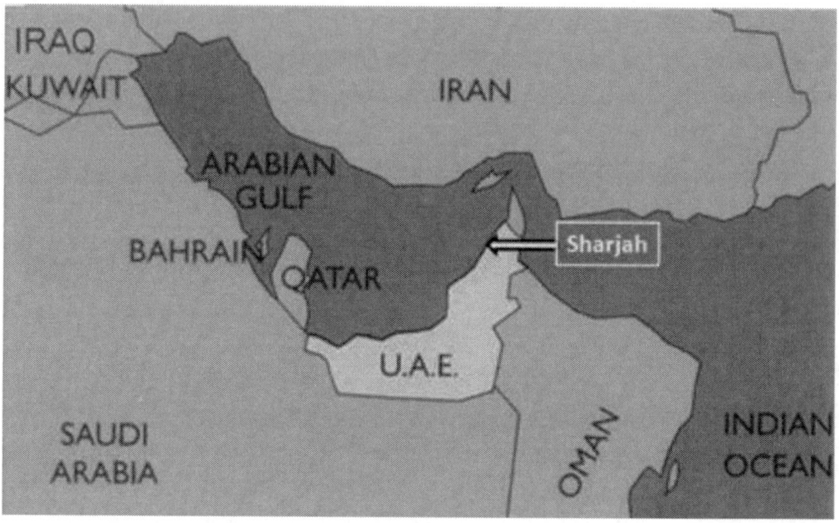

Figure 1. The location of the coastal city of Sharjah, UAE.

3. Social Survey

Over 1000 survey were distributed door to door, sent by mail or face-to-face interviews. The ages of respondents were restricted to 18 and above and respondents must have lived in their dwelling for more than one year. Questions were carefully designed following international guidelines and recommendations available in the literature [11,12,13], with slight additions that included Arabic translations of the questions and changing some of the questions to fit the research goals. Due to the bilingual nature of the city residents, the survey was conducted in Arabic and English, with the exact questions and wording. The survey is subdivided into two parts, the first of which contained demographic question such as age, gender, education level, district/zone, duration of residence in the dwelling and occupation. In the second part of the survey, respondents were asked questions aimed at gauging residents' annoyance level and feeling towards noise levels in and around their houses. For example, to determine the respondents' level of annoyance towards noise, we used a 10-point numeric scale that ranged from "*not at all annoyed*" to "*extremely annoyed*" (see Figure 2).

14. How annoying is transportation noise to you when you are at home? Please circle a number.

1	2	3	4	5	6	7	8	9	10
Not at all annoyed		Moderately annoyed		Annoyed		Very annoyed		Extremely annoyed	

14. ما هو تقييمك لمدى انزعاجك من أصوات المركبات الألية ؟

10	9	8	7	6	5	4	3	2	1
منزعج جدا		منزعج		منزعج نوعا ما		منزعج قليلا		غير منزعج تماما	

Figure 2. An example of a question in English and in Arabic.

The questionnaire also asked respondents questions on relevant issues such as the degree of noisiness of the acoustic environment in the respondents' house or apartment as perceived by them; and respondents personal noise sensitivity; residents' satisfaction with their living environment; the impact of transportation noise on their daily activities such as watching TV, conversation, sleep and concentration.

A total of 339 out of the 1000 surveys distributed were completed and returned, which gave us a response rate of 33.9%. Considering our past experience, this return rate is quite high, and this is due the social and culture trends where people are not used to such social survey and they do not give them much attention, especially since we did not offer any reward/incentives for filling the survey. In fact a number of respondents asked for such rewards, but we offered none. Despite this, the number of surveys return is considered the best that could be obtained given the circumstances and the results are expected to provide useful data on the annoyance response of residents to mixed traffic noise in a fast growing city that is witnessing an unprecedented economic growth and prosperity accompanied by an increase in noise pollution sources, especially transportation. It should be noted that the traffic on the roads does not include train noise since the city does not have train service.

4. RESULTS AND DISCUSSION

Respondents were 60% male and 40% female, the age distribution of which is shown in Table 1. The education level included high school (44.7%), BSc (48.6%), MSc and PhD (6.5%), while there was only (0.2%) were uneducated.

Among the residents interviewed 64% have lived in their dwelling for more than 3 years, and the rest (36%) of them have lived in their dwelling for a period between 1-3 years. 45% of the respondents live in houses, while the rest (55%) live in apartment blocks. The response to the question "*how far is the front of your residence from the road?*", 72% said they live < 50 m from the main road, while 28% live from 50 – 100 meters away from the road.

Table 1. Age distribution of respondents

Age	%
18 to 25	71.5
26 to 35	8.6
36 to 45	10.4
46 to 55	8
> 55	1.5

The response to the question: *how noisy is your neighborhood?* is shown in Figure 3. As shown in Figure 3, almost 50% of respondents felt that their neighborhood is noisy; with almost 30% feeling their neighborhood is very noist to extremely noisy. This was not a surprise to us since the measured noise level in these neighborhoods was high, exceeding 75dBA in many occasions [6]. On the annoyance question, see Figure 2, the answers of respondents are shown in Figure 4 below.

As shown in Figure 4 above, almost 50% of respondents felt annoyed, very annoyed and extremely annoyed because of the noise, while only 22% of the respondents felt not at all annoyed. As one may expect, the results in Figure 4 correlated very well with those in Figure 3, with r value of 0.98 and p=0.003.

We have also investigated the difference in the response of respondents for each individual district studied and we have observed a noticeable drop in the number of satisfied individuals (not at all annoyed) to 12% in one district, which is labeled as an industrial district and known for its heavy traffic volumes.

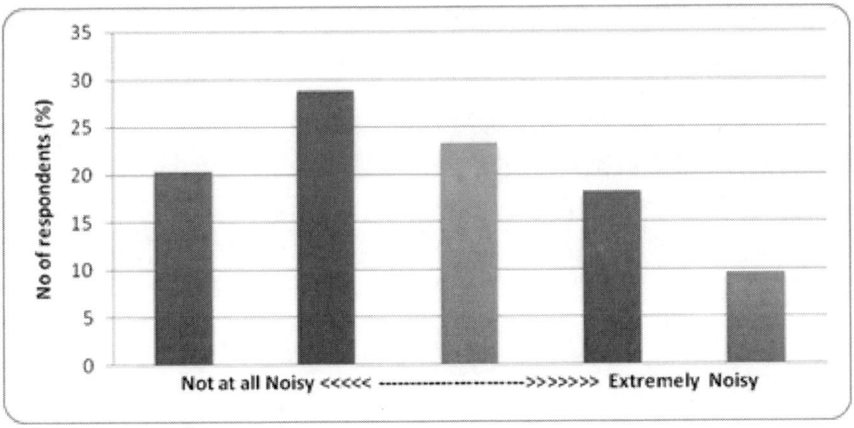

Figure 3. The response of respondents to the question on *how noisy is their neighbourhood?*.

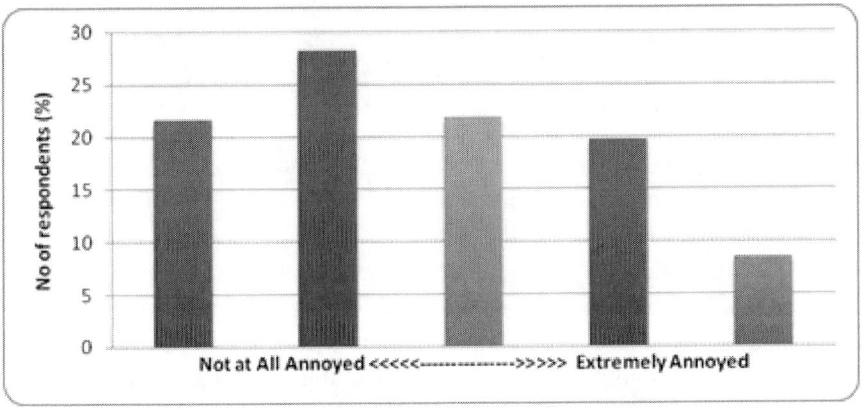

Figure 4. The response of respondents to the question on *how annoyed are you from trasportatio noise?*.

Table 2. The effect of noise on leisure activities: *does noise interfere with your leisure activities*?

Activity	Strongly disagree ←←——→→ Strongly agree					
Watching TV	1	2	3	4	5	6	7
Sleeping	1	2	3	4	5	6	7
Conversation	1	2	3	4	5	6	7
Concentrating	1	2	3	4	5	6	7

The last question, which was directed to respondents, was about transportation noise and its interference with leisure activities such as watching TV, sleeping, conversation and concentration.

As shown in Table 2, respondents were asked to scale the interference of noise with these activities. Subjects' responses are summarized in Figure 5.

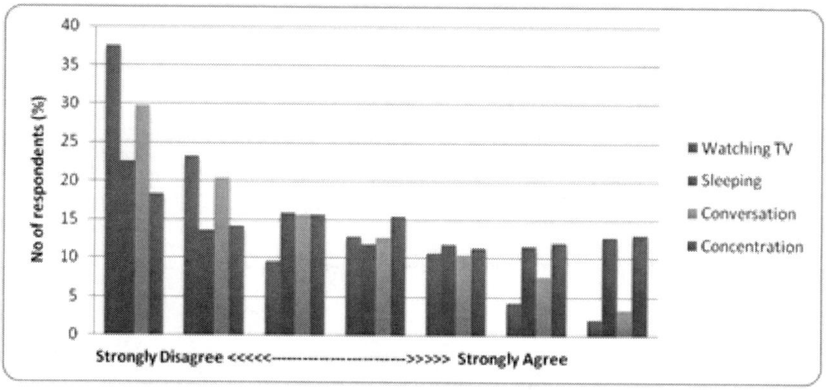

Figure 5. Noise interference with leisure actvities.

It is clear from Figure 5 sleeping and concentration are the most effected activity by noise among all leisure activities, with almost 30% of respondents gave it 5 or more on the agree-disagree scale. Watching TV and conversation are the least affected. It should be noted that we did not notice any gender preference in this question. It is also interesting to report that when residences asked if they have ever complained about noise, over 85% have never complained, which may be attributed to the fact that residents felt that noise is unavoidable and they have to live with it. Also, it was evident that a number of residents were not aware of the fact that exposure to high noise levels over long periods lead to adverse health effects.

The numbers of annoyed individuals in the city of Sharjah is considered high. In comparison to other studies performed in other countries in Europe, North America and Australia, where noise is perceived differently and considering the high noise levels, the number of annoyed individuals is expected to be higher [14]. This is true since noise is attributed to many factors, most of which are subjective and depend on personal attitude and how individuals perceive the noise. It is also found that annoyance is affected by demography, living condition, noise awareness, and social status may be a determinant factor in peoples' perception of noise [15].

CONCLUSION

The results presented in this paper indicate that environmental noise is a cause of concern in the city of Sharjah. Previous reports showed noise levels in the city are well over 65dBA in more than 75% of the districts studied. To determine the annoyance level, a social survey was carried out and has shown that noise is indeed a serious problem in the city of Sharjah. In particular it was found that almost 50% of the residents surveyed felt annoyed and that noise interferes with their sleep and concentration. This number is considered high and unacceptable. It is obvious that a number of administrative and technical measures needed to be taken in order to improve the noise situation in the city of Sharjah. It is left to future reports to correlate the percentage of highly annoyed to the noise levels in these districts. Such correlation will enable us to compare the results to similar studies performed in other countries.

ACKNOWLEDGMENT

The author is grateful to the College of Graduate Studies and Scientific Research, University of Sharjah, for providing funding for the project and thanks to the students who helped in distrusting and collecting the surveys.

REFERENCES

[1] G. H. Pandya. "Assessment of Traffic noise and its impact on the community". *J. Environ. Studies*, December Vol. 60(6), 595–602 (2003).

[2] Jarupet. al. (The HYENA Team). "Hypertension and Exposure to Noise Near Airports: the HYENA Study". *Environmental Health Perspectives*,. Vol 116(3): 329-333 (2008).

[3] Mats Wilhelmsson. "The Impact of Traffic Noise on the Values of Single-family Houses". *Journal of Environmental Planning and Management*, Volume 43, Issue 6 November, pages 799 – 815 (2000).

[4] W Passchier-Vermeer and W F Passchier. "Noise exposure and public health". *Environ Health Perspect.* March; 108 (Suppl 1): 123–131(2000).

[5] World Health Organization 1999 Guidelines for Community Noise.http://www.who.int/docstore/peh/noise/guidelines2.html (1999).

[6] H. M. Elmehdi, and N. M. Nayfeh. "Noise Levels near Roads of Major Cities of the United Arab Emirates Noise Levels near Roads of Major Cities of the United Arab Emirates". *The 19th International Congress on Acoustics* - ICA07MADRID, Spain, 2-5 September 2007.

[7] H.M.ELmehdi. "The relationship between civil aircraft noise and community annoyance near Dubai International Airport (UAE)" *Proceedings of ACOUSTICS 2008*.24-26 November 2008, Geelong, Australia (2008).

[8] UAE Interact. http://uaeinteract.com/news/default.asp?ID=320 (2008).

[9] ANDREW SMITH. "A review of the effects of noise on human performance". *Scandinavian Journal of Psychology*, Volume 30 Issue 3, Pages 185 – 206 (2008).

[10] H. N. Hansell. The behavioral effects of noise on man: the patient with "intensive care unit psychosis". *Heart Lung*. Jan;13(1):59-65 (1984).

[11] H.M. E. Miedema and H. Vos. "Exposure-response relationships for transportation noise".*J. Acoust. Soc. Am.* Volume 104, Issue 6, pp. 3432-3445 (December 1998).

[12] Fields, J. M., de Jong, R. G., Gjestland, T, Flindell, R. S., Job, F., Kurra, S, Lercher, P, Vallet, M., Yano, T., Guski, R., Felscher-suhr, U, and Schumer, R. "Standardized general-purpose noise reaction questions for community noise surveys: Research and a recommendation". *Journal of Sound and Vibration*. 242(4) 641-679 (2001).

[13] Schultz, T. "Synthesis of social surveys on noise annoyance". *J Acoust. Soc. Am.* (1987) 64(2): 377- 405.

[14] Kin-Che Lam et al. "Annoyance response to mixed transportation noise in Hong Kong". *Applied Acoustics* 70, 1–10 (2009).

[15] H.M. E. Miedema and H. Vos. "Demographicand attitudinal factors that modify annoyance from transportation noise". *J. Acoust. Soc. Am.* Volume 105, Issue 6, pp. 3336-3344 (June 1999).

In: Noise Pollution
Editor: Eugene Roberto Nicchi
ISBN: 978-1-63321-110-0

Chapter 10

NOISE POLLUTION IN ZIMBABWE: SOURCES, EFFECTS AND HEALTH IMPLICATIONS

Jemitias Mapira
Department of Geography and Environmental Science,
Great Zimbabwe University, Masvingo City, Zimbabwe

ABSTRACT

Noise pollution is a growing problem in Zimbabwe and other developing countries especially those in sub-Saharan Africa. It impairs people's hearing abilities, damages their health and can lead to death. This chapter examines noise pollution in Zimbabwe with special reference to sources, effects and health implications as well as work place productivity. It is based on primary and secondary sources of information. A questionnaire survey was conducted in December, 2013 in order to determine various aspects of the problem in both rural and urban areas of Zimbabwe. It was preceded by literature review on the subject matter. Some of the issues discussed include: the country's legislation on noise pollution, causes and effects of noise in urban and rural areas as well as suggested solutions on the problems and challenges exposed. It is noted that noise pollution is more pronounced in urban than rural areas as indicated by the number of victims.

The chapter ends with an examination of noise pollution in Harare's furniture manufacturing sector. It shows that levels of pollution in most cases exceed those prescribed by the country's laws on noise pollution.

Consequently, solutions should be found in order to address the problem at local and national levels. However, these solutions should be based on research if they have to be effective. Since Zimbabwe is a cash strapped economy, it lacks the capacity to fund such expensive research projects. There is a need for the country to solicit for foreign donations so as to conduct researches on noise pollution issues. However, before this can be done, the country will have to improve its relations with some Western donor countries, such as Britain and the USA.

INTRODUCTION

Noise pollution is a growing problem in most developing countries (Katyal and Satake, 2001) and Zimbabwe is not an exception. Although it is not as severe as it is in industrialized nations, it is a cause for concern due to its negative effects on human health and work place productivity. Noise has been defined as "*any unwanted, disturbing, or harmful sound that impairs or interferes with hearing, causing stress, hampers concentration and work efficiency, or causes accidents*" (Miller and Spoolman, 2009:596). Goodal (1987, 324) states that unwanted sound "*may be socially undesirable because it intrudes on, disturbs or annoys people, or medically undesirable-very high levels of sound can cause hearing damage*". In some cases, noise can be fatal. Noise pollution is generally more pronounced in urban than in rural areas. Urban dwellers are therefore more exposed to noise pollution than rural folks (Miller and Spoolman, 2009). This is due to the concentration of traffic flows, human, industrial and commercial activities in these areas. In rural areas, sources of noise pollution include: open cast mines, grinding mills, beer halls, schools, churches and shopping centers. The main sources of noise in towns and cities are: manufacturing industries, airports, seaports, city centers, beer halls, night clubs, churches, schools and sports stadia.

NOISE POLLUTION IN GENERAL

Noise levels are measured in decibels (dbA) which "*describe the un-controlled natural or random variation in a data set. It is used to explain... after the effects of known variables have been accounted for*" (Goodal, 1987: 324). According to Miller and Spoolman (2009), sound pressure (or noise) becomes harmful at around 85dB(A) and painful at about 120 db(A).In extreme cases such as 180 dB (A), it can kill (Katyal and Satake, 2001). If human beings are exposed to high sound levels (above 85dB(A), their hearing may be permanently impaired. In the USA, for example, about one of every eight children and teens have "*some permanent hearing loss, mostly from listening to music at loud levels*" (Miller and Spoolman, 2009:596). However, it should be noted that people react differently to sound pressure (or noise), as some are more resistant than others (Katyal and Satake, 2001).

In most industrialized countries, there are regulations, which govern noise pollution levels. For example, the European regulation for fireworks explosions is that noise levels should not exceed 120 decibels (Miller and Spoolman, 2009). Although noise pollution is a cause for concern, there is a dearth of literature on the subject especially in developing countries. This is due to the fact that researchers focus on other types of pollution such as: air and water, which are considered to be more harmful to human health (Miller, 1996). Hence, the need for research on this subject cannot be over-emphasized. This chapter seeks to fill this information gap through an examination of the noise pollution problem in Zimbabwe. The discussion is based on information derived from existing literature and a questionnaire survey, which was conducted in various parts of Zimbabwe in December, 2013. Issues included in the chapter are: sources, effects, implications on health, and possible solutions to the problem.

Zimbabwe's Legislation on Noise Pollution

According to its last population census, Zimbabwe has a total of 12.9 million people (CSO, 2012). Although it is a developing country, it has a fairly sophisticated economy, which comprises: agriculture, mining and manufacturing (Munowenyu, 1996). The country is endowed with over 40 minerals, which have become the backbone of its economy and a source of raw materials for the manufacturing sector. During the 1990s, "*the mining industry employed over 60 000 people, generated some 45% of the country's foreign currency and also contributed significantly to the nation's manufacturing industry*" (Mapira and Zhou, 2006: 48). In spite of this valuable contribution to the national economy, mining is notorious for causing land degradation, air, water and noise pollution (Munezvenyu, 1992). The mining industry is governed by the Mines and Minerals Act of 1996 and the Environmental Management Act of 2002. The former Act was established during the colonial era in 1961 and has been revised many times since then.

In 2007, an Environmental Management Agency (EMA), was established through the merging of three government departments (G.o.Z, 2009), which included: the Natural Resources Board (NRB), the Zimbabwe National Water Authority's (ZINWA) Water Quality Section, and the Ministry of Health and Child Welfare's Hazardous Substances and Atmospheric Pollution Sections. EMA falls under the Ministry of Environment and Natural Resources Management (G.o.Z, 2009). It was established through Statutory Instrument 103 of 2007 of the Environmental Management Act Chapter 20:27 (G.o.Z, 2009). It is the country's environmental protection and management agency (Chimhowu, Manjengwa and Feresu, 2010). In order to execute its mandate, EMA is empowered to perform several tasks. Firstly, it sets minimum standards to be prescribed for the emissions of noise and vibration. Secondly, it prescribes criteria and procedures for the measurement of noise and vibration pollution into the environment. Thirdly, it establishes criteria and procedures to be prescribed for the measurement of sub-sonic vibration.

Furthermore, it sets noise level and noise emission standards that are applicable to: construction sites, plant machinery, motor vehicles, aircraft, industrial, and commercial activities. Finally, it prescribes guidelines for the abatement of unreasonable noise and vibration pollution emitted into the environment from any source. Section 80 of the Environmental Management Act (Chapter 20: 27, 385), states that "*Any person who emits noise in excess of the noise emission standards prescribed in terms of section seventy-nine shall be liable upon conviction to a fine not exceeding level fourteen or to imprisonment for a period not exceeding twelve months or to both such fine and such imprison-ment*".

However, there are two exceptions in respect of stipulated noise standards, according to Section 81. Firstly, notwithstanding Section eighty, EMA can grant a temporary permit, on request, not exceeding three months, allowing emission of noise in excess of prescribed standards for such activities as fireworks, demolitions, firing ranges and specific heavy industry undertakings on such terms and conditions as the agency may determine. Secondly, where an exemption has been granted under subsection (1), workers exposed to excessive levels of noise should be adequately protected in accordance with prescribed measures or the directions issued by the agency.

PROTECTION AGAINST NOISE AT THE WORKPLACE

In Zimbabwe, the Factories and Works (General) Regulations of 1976 seek to promote health and safety at work places. Consequently, noise pollution is a crime just as it is in other countries. While in Europe, the accepted noise level for fireworks is 120 dB (A), in Zimbabwe the figure (90 dB (A) is comparatively low. This implies that noise pollution restriction laws in the country are stricter than they are in Europe. However, since sound pressure becomes harmful from 85 dB (A) onwards, the country's law can be regarded as rather lenient. Section 6 of the Act (pages 1112 to1113) states that:

1. *"No person shall be exposed to sound-levels exceeding the limits prescribed in subsections 2, 3 and 4 of this section unless such a person has been supplied with, and is using, ear-protectors of a type approved by an inspector, which reduce the sound-level reaching the user's ear to or below the limits allowed for an unprotected ear.*
2. *If exposure to sound-levels is continued for eight hours in any one day and is to a reasonably constant volume of sound, the sound-level shall not exceed 90 dB (A): Provided that (i) if exposure is for periods other than eight hours, or if the sound-level is fluctuating, an equivalent continuous sound-level (herein referred to as Leq) shall be calculated according to the prescribed formula, and this value shall not exceed the equivalent of 90dB (A); (ii) if the non-continuous exposure cannot be adequately measured and controlled, any exposure at a sound-level in excess of 90 dB (A) shall be regarded as exceeding 90 dB (A).*
3. *Under no circumstances should the unprotected ear be exposed to a sound-pressure-level measured with an instrument set to the 'fast' response exceeding 135 dB (A) or in the case of impulse noise, an instantaneous sound-pressure exceeding 150 dB (A)*
4. *Other parts of the body should not be exposed to a sound-pressure-level measured with an instrument set to the 'fast' response exceeding 150 dB (A)*
5. *Areas where persons may be exposed to sound-levels exceeding the limits set out in this section shall be identified as ear-protection areas and the boundaries shall be clearly defined*
6. *Entry to ear-protection areas must be restricted to those authorized to do so and all such persons must use effective ear-protection*

7. *A prominent warning notice bearing unauthorized entry, and entry without the use of ear protection must be posted near every entrance to an ear protection area*
8. *Machines which in normal operations are likely to produce a sound level at the operator's ear, exceeding the limits set out in this section, must carry a prominent warning notice which shall be displayed in a position clearly visible to the operator".*

In the past, Zimbabwe has been criticized for failing to take environmental issues seriously (Lopes, 1996) Although its environmental laws are well documented, they are not always implemented at the grass roots level (Chimhowu, et al., 2010). For example, the country's legislation on noise pollution dates back to the colonial era (1976). However, to date few or no offenders have been convicted, indicating apathy at the law-enforcement level. One of the reasons is that noise pollution is not considered as a serious problem compared to other types of pollution such as those of air and water. Even at global level, there is more literature on air and water contamination than that on noise pollution (Katyal and Satake, 2001). This is in spite of the severe damage it has on people's hearing abilities (Miller and Spoolman, 2009).

Noise Pollution in Zimbabwe's Urban Areas

According to a national survey, conducted in December 2013, the main sources of noise pollution in Zimbabwe include: vehicular traffic, aircrafts, beer halls and night clubs, churches, industries, construction work, railway trains, shopping centers, squatter settlements, street kids, radios and TVs, in that order (Table 1). The figures are derived from the number of respondents who mentioned each source of noise pollution (such as vehicular traffic). These figures were later arranged according to their sizes in order to rank them. People who live close to these sources are exposed to noise pollution and they risk being negatively affected in one way or another. Although people react differently to noise pollution, most cannot endure it for a long time (Katyal and Satake, 2001).

Table 1. Main Sources of Noise in Zimbabwe

Source	No of Respondents	Ranking
Vehicular traffic	67	1
Aircraft	48	2
Beer halls and night clubs	47	3
Churches	36	4
Industries	36	4
Construction work	28	6
Railway trains	28	6
Shopping centres	26	8
Squatter settlements	23	9
Street kids	21	10
Radios and TVs	18	11

Source: Questionnaire Survey.

In the city of Masvingo, the survey identified ten sources of noise (Table 2). According to the table, the most frequently cited sources are: powere generators, beer halls and night clubs, industries, radios, moving road vehicles, churches, shopping centers, railway trains, panel beastyers and road shows. Diesel power generators are numerous due to the erratic supply of electricity in the city (Mapira and Munthali, 2011). Although only a minority of households, own some generators, they pose a serious threat to the tranquility of their residential areas. People who live close to places of entertainment such as beer halls and night clubs also have to endure a lot of noise pollution just as those who reside near industrial sites. The SIMBI Iron Ore Processing Plant in the city is also a notorious source of noise pollution.

Its premises are occupied by numerous vehicles, which generate noise (Table 3). According to a recent publication, the main sources of noise pollution at the iron and steel manufacturing plant include: "*compressors, vibrating screens, pumps, power generators, earth moving vehicles such as loaders, cranes and dumpers. Although to date apparently no victims of ...noise pollution are known to have been hospitalized, the local people's continual exposure to this problem is a health hazard, which requires urgent solutions*" Mapira (2012b, 181).

Table 2. The Main Sources of Noise Pollution in the City of Masvingo

Source	No. of Respondents	Ranking
Power generators	52	1
Beer halls and night clubs	43	2
Industries	37	3
Radios	35	4
Moving road vehicles	35	4
Churches	33	6
Shopping centers	27	7
Railway trains	26	8
Panel beaters	24	9
Road shows	21	10

Source: Questionnaire Survey.

CAUSES AND EFFECTS OF NOISE POLLUTION IN URBAN CENTERS

Noise pollution affects people in various ways. According to a recent survey, it irritates their ears, disturbs conversations, impairs hearing abilities, disturbs sleep especially at night, retards communication during discussions and reduces individual concentration and attention. Pupils and students are also negatively affected by noise as it disturbs their lessons, studies, reading and learning in general. It also disrupts learning activities and consequently undermines effective study.

Table 3. Numbers of Vehicles at SIMBI PVT Ltd Iron and Steel Manufacturing Plant

Type of Vehicle	Fleet Size
Motor Cycles	18
Motor Cars	12
Heavy 30 Tone Trucks	6
Graders	4
Excavators	4
Dumpers	4
Lorries	3
Front end loaders	3

Type of Vehicle	Fleet Size
Buses	2
Rock drillers	2
Tippers	2
Cranes	1
Total	61

Source: Mapira (2012b:180).

Among workers and employees noise is also a nuisance as it disturbs normal working, reduces concentration and productivity, causes accidents due to loss of concentration, creates communication barriers, and is time wasting as it interferes with communication among employees. Elderly citizens also experience a number of problems due to noise pollution. These include: impairing hearing abilities, disturbing sleep at night and during the day, irritating ears and damaging ear drums. To sick people, it is even more damaging as it is responsible for the following problems: it deprives them of the much needed rest, irritates them and worsens their sickness, robs them of a peaceful environment in which they can recover, failure to rest disturbs the healing process, and ultimately, it shortens their lives.

VICTIMS OF NOISE POLLUTION IN ZIMBABWE'S URBAN CENTERS

In keeping with global trends, Zimbabwe's urban centers are notorious for generating noise pollution whose effects have already been discussed. Table 4 depicts the victims in some of the country's towns and cities. Harare, Bulawayo, Kwekwe, Zvishavane, Mutare, Gweru, Masvingo,Chiredzi and Kadoma are major industrial centers in the country (Munowenyu, 1996). Apart from causing air and water pollution, they are also notorious for generating noise pollution. Zvishavane is an asbestos mining town while Chiredzi is well known for its sugar processing mills. On the other hand, before 2004, when SIMBI Pvt. Ltd was established, Masvingo was a relatively quiet city dominated by light industries such as packaging, butcheries, bottling and warehousing, which do not generate much noise (Mapira, 2012b). However, the opening up of an iron and steel manufacturing plant led to an upsurge in noise pollution at its site some 5km along the Bulawayo road.

Although no measurements on sound pressure levels were taken during the survey, the number of victims reflects the severity of the problem in all towns and cities. Some 364 victims comprising men, women and children were identified as victims of noise pollution. These people have lost their hearing abilities and cannot cope without hearing devices. While some have been hospitalized, others have joined institutions of the deaf such Jairos Jiri and Copota School of the Deaf and Dumb. It is pertinent to note that males outnumber their female counterparts remarkably. This probably indicates that males are more exposed to noise pollution than females. In general, more men work in heavy industries, construction work and spend more time in beer halls and night clubs than their female counterparts who are usually confined to their homes.

Table 4. Victims of Noise Pollution in Zimbabwe's Urban Centers

Town or city	Male victims	Female victims	Totals
Harare	44	28	72
Bulawayo	37	14	51
Mutare	23	21	44
Kwekwe	28	15	43
Zvishavane	24	12	36
Chiredzi	18	16	34
Masvingo	18	13	31
Gweru	18	12	30
Kadoma	17	16	23
Grand Totals	226	138	364

Source: Questionnaire Survey.

SUGGESTED SOLUTIONS FOR URBAN NOISE POLLUTION

During the survey, urban dwellers suggested several possible solutions to the problem of noise pollution. Firstly, there is a need to locate heavy industries far away from residential, social and institutional land-uses so that people are not exposed to the dangers of noise pollution.

Secondly, sound conditioning devices should be installed in cars, buses and machines at work places so as to insulate people against noise. Thirdly, laws should be enacted so as to restrict radio and TV volumes.

Furthermore, motor vehicles should be fitted with silencing devices while airports, beer halls, night clubs and churches should be located far away from residential areas where they expose people to excessive noise levels.

Finally, people should be educated about the dangers of noise pollution so that they can take necessary precautions. This can be achieved through anti-noise pollution campaigns conducted in towns and cities.

However, in order for these strategies to yield fruit, they should be backed by the necessary political will among policy makers. As mentioned previously, Zimbabwe has, in the past, failed to take its environmental laws seriously (Lopes, 1996). For example, although environmental impact assessments (EIAs) are mandatory, they have not always been implemented (Chimhowu, et al., 2010). This is because political and economic considerations usually take precedence over environmental concerns (Chiwota and Hauge in Lopes, 1996).

NOISE POLLUTION IN RURAL AREAS

Although Zimbabwean rural areas are generally characterized by peace and tranquility, noise pollution is also common in some places. The main sources of noise include: grinding mills, open cast mines, road traffic, moving trains, beer parties, shopping centers, livestock and wild animals such as hyenas, foxes and baboons (Table 5).

Table 5. The main sources of noise in rural areas

Source	Respondents	Ranking
Grinding mills	92	1
Open cast mines	67	2
Road traffic	53	3
Moving trains	46	4
Beer parties	42	5
Shopping centers	41	6
Livestock (cattle, donkeys and goats)	40	7
Wild animals (hyenas, foxes and baboons)	38	8
Schools	35	9
Open air churches	31	10
Traditional ceremonies	31	10
Radios	30	12

Source: Questionnaire Survey.

Other sources include: schools, open air churches, traditional ceremonies and radios. People who live close to grinding mills, open cast mines, roads and railway lines have to endure a lot of noise pollution. Beer parties, shopping centers, livestock and wild animals are also notorious for noise pollution. Other sources include: schools, open air churches and traditional ceremonies, which are usually accompanied by drum beats. However, compared to urban areas, rural communities are far less exposed to noise than urban dwellers.

EFFECTS OF NOISE ON RURAL PEOPLE

Rural folks are negatively affected by noise in various ways including: loss of concentration, memory loss, lack of peace and tranquility, worsening illness, causing short tempers, tiredness, heart problems, and sleeplessness. Noise also disturbs the working spirit thereby undermining productivity at the work place. This study discovered that different categories of people in rural areas do not react in the same ways to noise pollution. They include: ordinary villagers, pupils and students, workers and employees, elderly people and the sick (Table 6). Ordinary villagers who are exposed to noise experience problems such as: headaches, irritation, communication barriers and impaired hearing abilities. Other challenges include: disturbed thought processes, sleeplessness, damaged ear drums and irritated nerves.

Noise pollution is also a threat to education as pupils and students experience problems such as: reduced attention and concentration, disturbed studies and thought processes as well as reduced school work performance.

Workers and employees also face numerous challenges when exposed to noise at their work places. Reduced concentration undermines goal achievement while increasing the possibility of accidents. Hearing problems become rampant among workers while tiredness leads to mistakes, which reduce both quality and productivity. Efficiency is undermined while the economy suffers in the long run. To elderly people, noise is also a nuisance as it causes confusion, impairs hearing, irritates ears and causes sleeplessness.

Ultimately, it is a threat to their health. In the case of those who are sick, it can be fatal. This is because it worsens pain or illness, disturbs rest or sleep, irritates ears, causes heart problems and may lead to death.

Table 6. Effects of noise pollution on different categories of people

Category	Effects
Ordinary villagers	Causes headaches, pollutes ears, hinders communication between people, impairs hearing abilities, disturbs thought processes, causes sleeplessness, damages ear drums and irritates the mind.
Pupils and students	Disturbs attention and concentration, studies and thought processes. Also undermines school work performance.
Workers and employees	Reduces goal achievement, can lead to accidents at work places. Reduces concentration at work, causes hearing problems and undermines productivity and efficiency at workplaces. Causes tiredness, which may lead to mistakes which undermine quality.
Elderly people	Causes mental confusion, impairs hearing, causes sleeplessness and irritates ears.
Sick people	Worsens pain or illness, disturbs rest or sleep, irritates ears and causes death. Also causes heart problems.

Source: Questionnaire Survey.

VICTIMS OF NOISE POLLUTION IN ZIMBABWE'S RURAL AREAS

There are fewer victims of noise pollution in rural areas compared to towns and cities (Table 7). While urban areas recorded some 364 victims (Table 4), their rural counterparts accounted for only 131. This probably reflects differences in the geographical distribution of noise pollution in the country. In general, noise pollution is more pronounced in urban areas compared to rural areas (Goodal, 1987). Heavy industries, road traffic and social gatherings are far more common in towns and cities than in rural areas. These activities generate a lot of noise that is rarely found in rural areas. Victims of noise in rural areas range from the age of 15 to 65 and cut across both sexes even though males still outnumber females just as in the case of towns and cities (Table 4).

Suggested Solutions for Noise Pollution

Rural folks also suggested several solutions to the problem of noise. They included:

a) The need for tough legislation and stiff penalties on offenders
b) Educating the public on the dangers of noise pollution
c) Beer halls and night clubs should regulate their noise or risk prosecution
d) Grinding mills and other sources of noise should be located far away from residential areas
e) Heavy vehicles should use roads which are far away from residential areas
f) Mines and quarries should not be too close to people's homes
g) Efforts should be made to minimize noise at work places
h) Government should have a more binding policy on noise pollution
i) Public campaigns against noise pollution should be conducted throughout the country
j) Diesel grinding mills and trains should be replaced by electir ones
k) People should wear ear protectors when they venture into noisy areas
l) Industries should also develop technology that reduces noise pollution in both rural and urban areas
m) Domesticated animals should be kept far away from residential areas
n) Noise from radios and TVs should be reduced
o) Rooms and buildings should be insulated against noise pollution, and
p) Road vehicles should be fitted with silencers so that their noise is reduced.

Government departments such as the Environmental Management Agency (EMA) can play a pivotal role in the implementation of some of these solutions. This is because without backing from policy makers, little or no progress in this direction can be expected, as mentioned previously.

Table 7. Victims of Noise Pollution in Rural Areas

Area	Male victims	Female Victim	Total
Bindura	10	4	14
Buhera	9	11	20
Chimanimani	8	6	14
Chinhoi	7	7	14
Chipinge Rural	6	5	11
Chivhu	7	7	14
Gwanda	4	3	7
Lupani	3	2	5
Mberengwa	5	4	9
Zaka	7	5	12
Zvishavane Rural	6	4	10
Grand Total	73	58	131

Source: Questionnaire Survey.

NOISE POLLUTION IN HARARE'S FURNITURE MANUFACTURING INDUSTRY

A major theme, which runs through this study is that of workers' health and their productivity. This is examined in this section. Mutetwa and Dozva (2011) conducted a survey, which was aimed at investigating noise exposures within Harare's Manufacturing Industry. The furniture manufacturing sector plays a key role in Zimbabwe's economy. In the city of Harare some 400 workers are employed. Due to the technology used, workers are exposed to noise levels, which may damage their health and reduce their productivity at the work place.

Some twelve companies were targeted and the objectives of the survey were to: identify furniture manufacturing companies with high levels of noise pollution, determine whether these companies have hearing loss prevention programs, and recommend appropriate controls to mitigate against possible noise induced hearing loss. Results of the survey for cross-cutting operations, indicate that noise exposure levels range from 92.6 dB (A) to 110.7 dB (A), with an average of 99.6 dB (A). In spindle operations, they range from 94 dB (A) to 104.7 dB (A) with an average of 98.5 dB (A). In the case of drum sanding operations, exposure levels range from 91.5 dB (A) to 99 dB (A) with average exposure levels being 94.4 dB (A).

In orbital sanding, noise exposure levels range from 88 dB (A) to 99.4 dB (A), while they average 94 dB (A). Routering machine operations record levels which range from 84.3 dB (A) to 93.7 dB (A) with an average exposure level of 89.9 dB (A). In the case of molding machine operations, noise exposure levels range from 89 d B (A) to 101.4 dB (A) while the average is 97.8 dB (A). In the polishing workshops, noise exposure levels range from89.2 dB (A) to 97.7 dB (A) with an average exposure level of 86.3 dB (A). On the other hand, in spray rooms, levels of exposure range from 74.4 dB (A) to 93.5 dB (A) while average exposure was 86.3 dB (A). In chip board operations, the researchers noted that noise exposure levels lay in the range of 85.5 dB (A) and 96.7 dB (A) while the average exposure was 91 dB (A).

In the belt sanding operations, exposure levels are in the range of 84 dB (A) to 100.4 dB (A). On the other hand, the average exposure level was 91 dB (A). Ripping operations recorded exposure levels, which range from 90.4 dB (A) to 103.2 dB (A) with 97.7 as the average level. In the final fitting areas/upholster departments, the type of noise is mostly impulsive in nature. Here exposure levels range from 92.3 dB (A) to 131.5 dB (A) while the average exposure level is 100.7 dB (A). Finally, in the thicknesser/planning machine operations, noise exposure levels range from 90.3 dB (A) while average exposure level was 98.1d B (A).

After providing noise exposure levels for all the targeted departments, Mutetwa and Dozva (2011, 3) conclude that "*noise exposure levels in the furniture manufacturing industry in Harare exceeded the acceptable exposure limit of 85 dB (A) for continuous noise which is the set legal standard in most countries. Therefore workers in this sector are at an increased risk of hearing loss. With regard to impulsive noise, exposure levels were observed to be within the acceptable limit of 150 dB (A) according to Zimbabwe Factories and Works regulations as well as being within the 140 dB (A() limit which is the set upper limit legal standard for peak sound pressure level in most countries. It is therefore recommended that companies evaluated should, without delay, establish and implement a comprehensive hearing loss prevention programme*". This inevitably calls for funds, which are scarce in most companies.

The Government of Zimbabwe does not always take environmental issues seriously as Lopes (1996) and Chimhowu, et al. (2010) claim. This is a serious weakness, which should be addressed as soon as possible. However, in some cases, the problem has been the lack of funds rather than sheer apathy among policy makers (Mapira, 2012a). From this perspective, donor funding could be a possible solution. However, the country would have to improve its international relations with some Western countries if it has to attract foreign aid (Mapira and Munthali, 2011). But this is likely to take time as some vigorous lobbying will be required in order to achieve this goal.

CONCLUSION

This chapter has examined the problem of noise pollution in Zimbabwe with a view to making some recommendations at both local and national levels. It has discussed the problem in rural and urban areas. Although the problem is severe in urban areas due to the numerous noise generating activities, rural areas are also at risk to a remarkable degree. Existing literature shows that in Zimbabwe, just as in other developing countries, noise pollution has been a neglected area of research even though it has negative impacts on human health and work place productivity. This scenario has led to a dearth in noise pollution literature, an issue that should be addressed urgently. This can be in the form of sponsorship of such research efforts from developed countries. The fact that some 364 urban and 131 rural victims of noise pollution have been identified through a questionnaire survey should be viewed as a tip of the ice berg, that should stimulate more in-depth studies especially directed at finding workable solutions for the problem.

Due to strained budgets, countries in the developing world often lack funds to engage in such research. Noise pollution is an issue which calls for donor funding if thorough researches have to be done. Zimbabwe does not lack research experts on the subject. However, it is the funding that is required in order to execute the tasks. Mutetwa and Dozva's (2011) paper is an example of a well-researched study on the subject of noise pollution. If their research efforts could be extended to other aspects of the country's economy, more information could be generated for the benefit of researchers and policy makers. It should be borne in mind that without adequate, well researched information, policy implementation (at any level) is a futile exercise. The country should solicit for donor funding so that it can conduct detailed surveys on noise pollution at both local and national levels. However, there is also a need for the country to mend its relations with Western nations, which are the main sources of such funds (Chimhowu, et al., 2010).

REFERENCES

Central Statistical Office (CSO), *Census 2012 Zimbabwe Preliminary Report*, Harare.

Chimhowu, A. Manjengwa, J. and Feresu, S. 2010. *Moving Forward in Zimbabwe: Reducing Poverty and Promoting Growth*, IES, Harare.

Environmental Management Act (2002) Chapter 20: 27, Government Printer, Harare.

Environmental Management Agency (EMA), 2007, Government Printer, Harare.

Factories and Works (General) Regulations, 1976, Government Printer, Harare.

Goodall, B. 1987. *The Penguin Dictionary of Human Geography*, Penguin Books, London.

G.o. Z., 2009. *National Environmental Policy and Strategies*, Government of Zimbabwe, Ministry of Environment and Natural Resources Management, Harare.

Katyal, T. and Satake, M. 2001. *Environmental Pollution*, Anmol Publications Pvt, New Delhi.

Lopes, C. 1996. *Balancing Rocks: Environment and Development in Zimbabwe*, SAPES, Harare.

Mapira, J. and Zhou, T. M. 2006. 'From a Mining Settlement to a Police Camp: Buchwa Mine's Quest for Survival', *Zimbabwe Journal of Geographical Research,* Vol. 1 No. 1: 48-59.

Mapira, J. and Munthali, A. 2011. 'Household Enrgy Demand: Wood fuel Consumption and Peri-Urban Deforestation in the city of Masvingo (Zimbabwe)', *Journal of Sustainable Development in Africa* Vol. 13 No. 5: 264-279.

Mapira, J. 2012a. 'Zimbabwe's Environmental Education Policy and the Quest for Sustainable Development' *Journal of Sustainable Development in Africa,* Vol. 14 No. 6: 195-208.

Mapira, J. 2012b. 'Masvingo City's SIMBI Iron Ore Processing Plant: Socio = economic Benefits versus Environmental Threats' *Journal of Social Development in Africa* Vol. 27 No. 2: 165-188.

Miller, T. 1996. *Living in the Environment: Principles, Connections and Solutions*, Wadsworth Publishing Company, Belmont.

Miller, T. G. and Spoolman, S. 2009. *Introduction to Environmental Science*, Cengage Learning, Delhi.

Munezvenyu, P. 1992. 'Mining and its Effects on the Environment', *Geographical Education Magazine* Vol. 15 No. 1:24-29.

Munowenyu, E. M. 1996. *'A' Level Geography: A Comprehensive Guide*, Longman Zimbabwe, Harare.

Mutetwa, B. and Dozva, R. 2011. 'A Survey of Noise Exposures in the Furniture Manufacturing Industry in Harare, Zimbabwe', *On Guard: An Aid to Occupational Safety and Health* Vol. 17 June Issue No. 1: 3-10.

INDEX

A

B

C

D

E

I

J

L

M

N

O

S

T

U

V

W

Y

Z